AMERICAN PHILOSOPHICAL QUARTERLY

MONOGRAPH SERIES

AMERICAN PHILOSOPHICAL
QUARTERLY
MONOGRAPH SERIES

Edited by Nicholas Rescher

# STUDIES IN MODALITY

by
NICHOLAS RESCHER

with the collaboration of
RUTH MANOR    ARNOLD VANDER NAT    ZANE PARKS

Monograph No. 8    Oxford, 1974

PUBLISHED BY BASIL BLACKWELL

ISBN 0 631 11520 x

Library of Congress Catalog
Card No.: 73-80196

PRINTED IN ENGLAND
by Tinling (1973) Ltd., Prescot, Lancs
(a member of the Oxley Printing Group Ltd)

## CONTENTS

## EDITOR'S PREFACE

A word of apology regarding this monograph is in order. Throughout the initial decade of its history, it has been my systematic policy as editor of the *American Philosophical Quarterly* to exclude my own productions from its pages. The present monograph does not put an end to this intrinsically desirable policy of an embargo upon the editor's own work, but rather represents only a momentary lapse in its operation, by way of a grudging concession to recalcitrant circumstance. Plans for another publication collapsed in the last minute, and, rather than skipping a year in the sequence of monographs, the present publication was reluctantly decided upon.

## PREFACE

As regards developments in the philosophical sectors of logic, modal logic—the formal theory of reasoning with statements that involve a reference to possibility and necessity—has dominated the scene since World War II, and it is at present the most active and influential part of this problem-area. But although the subject itself has been advancing at a rapid rate in its abstract development, the exploration of its specifically philosophical implications and applications is comparatively underdeveloped. My own recent work in the field has largely run along these particular tracks. For one thing, I have sought to explore by way of historical case studies how central concepts of modal logic ideas have been involved in the work of various major philosophers. The five essays in Part 1 present case studies of this nature. All of them deal with an exposition of how some particular philosopher or school of philosophy has come to grips with modal conceptions and has proceeded to treat them in the development of their own ideas.

The first of these essays ("A New Approach to Aristotle's Apodeictic Syllogisms") seeks to maintain that the characteristic doctrine of Aristotle's modal syllogistic—the thesis that necessary conclusions can emerge in syllogisms with only one necessary premiss—rests on a basis that is ultimately not a matter of strict logic alone, but is grounded in metaphysical considerations regarding the structure of scientific explanation. The second essay ("The Theory of Modal Syllogistic in Medieval Arabic Philosophy") examines the details of a complex organon of modal reasoning based on purely tense-logical ideas, where modal statements merely represent theses as to the times at which certain circumstances obtain in this, the actual world, so that modalized contentions bear a purely actualistic orientation which implements the Aristotelian idea that "actuality is prior to possibility." The third essay ("Leibniz and the Evaluation of Worlds") considers the mechanisms by which the father of the theory of possible worlds proposed to deploy normative, value-theoretic considerations to distinguish the actual world from its "merely possible" alternatives. The fourth essay ("Kant and the 'Special Constitution' of Man's Mind") maintains that the conception of necessity operative in Kant's Critical Philosophy, in relation to the "universal and necessary" status of *a priori* synthetic truths, is not an *absolute* but a *relative* necessity, one whose force is relativized to certain essentially contingent features of the human cognitive faculties. The final essay of this part ("Bertrand Russell and Modal Logic") considers how one very influential 20th century thinker was led, on philosophical grounds, to take a negative stance toward modal conceptions, and to resist the development of a specifically modal logic. All of these cases share this common feature, that they exhibit the intimate linkage and consequence-laden interaction between the logician's purely philosophical positions on the one hand, and his doctrines regarding modal logic on the other.

On the nonhistorical side, I have been concerned with studying systematically some of the philosophical uses to which key ideas of the logic of modality can be put in application to philosophical issues of current interest. The four essays of Part 2 fall into this category. All of them address themselves to issues of epistemic logic and deal with the use of modal mechanisms in the construction of a formalized theory of knowledge.

The first of these essays ("On Alternatives in Epistemic Logic") argues that there is no one correct or optimal system in the formal logic of knowledge, but that the situation is inherently pluralistic, with a variety of vastly different logical systems reflecting, in a highly sensitive way, often very small differences in our assumptions regarding the mode of knowledge at issue. The second essay ("Epistemic Modal Categories and the Theory of Plausibility") constructs a mathematical, *quantitative* theory of modality—a theory analogous on the side of modal logic to the theory of probability on that of many-valued logic —to yield a tool for handling certain epistemic decision-problems. The third essay ("Restricted Inference and Inferential Myopia in Epistemic Logic") develops the logical machinery needed to implement the idea of certain systems of epistemic logic that the logical horizons of the knowing subject are limited, and that a knower may be postulated to know not *all* of the consequences of what he knows, but only those which are relatively "immediate." The final essay of this section ("Modal Elaborations of Propositional Logics and Their Epistemic Aspect") develops the machinery for systematizing an epistemic logic based on the thesis of logical omniscience, in that some *base*-system is presupposed, and a modalized meta-system constructed around this, subject to the supposition that a knower knows *everything* that can be established within the initial base-system. Throughout all of these studies, then, we are dealing with issues of the *application* of modal logic to the study of epistemological problems. All of them are concerned to develop logical mechanisms for studying issues in the theory of knowledge in a precise and formalized way.

All these essays are thus concerned with the *application* of the machinery of modal logic for the resolution of philosophical issues—some by way of a study of historical examples, others by way of bringing the tools of modal logic to bear on the clarification of current issues in the theory of knowledge.

Several of the studies comprising this book represent improved re-workings of previously published articles. (For the details see the citations given in the footnotes.) However, a simple reprinting of a prior publication is never at issue, in all cases the material has been revised and expanded for publication here.

Several of the studies are the products of collaboration: no's II and VI with Dr. Arnold vander Nat, no's I and VIII with Dr. Zane Parks, and No. IX with Dr. Ruth Manor. It gives me pleasure to acknowledge the assistance of these collaborators, and to thank them for agreeing to the inclusion of this material in the present volume.

# Part 1:
# HISTORICAL STUDIES

# I.

# A New Approach to Aristotle's Apodeictic Syllogisms

## 1. Introduction

VIRTUALLY all modal logicians after Aristotle have been troubled by his insistence that, given a valid *first figure* categorical syllogism (of the purely assertoric type *XXX*)

$$\frac{\begin{array}{c}\text{Major premiss } (P_M)\\ \text{Minor premiss } (P_m)\end{array}}{\text{Conclusion } (C)}$$

the corresponding modal syllogism (of type *LXL*) inferring the necessity of the conclusion from that of the major alone

$$\frac{\begin{array}{l}\text{Necessarily: } P_M\\ \quad P_m\end{array}}{\text{Necessarily: } C}$$

must be valid. The correspondingly *LLL* syllogism must, of course, also be valid *a fortiori*, while the corresponding *XLL* syllogism will—so Aristotle has it—be invalid. Despite extensive discussions of the problem, a convincing rationale for Aristotle's theory has yet to be provided.[1] The aim of the present discussion is to propose a suggestion along these lines.

The leading idea of the present proposal is that given syllogistic terms $\alpha$ and $\beta$ it is possible to define yet another term $[\alpha\beta]$ to represent the $\beta$-species of $\alpha$. As will be seen below, these bracketed terms represent a version of Aristotle's process of *ecthesis* ("selecting out" a part of the range of a syllogistic term). The $[\alpha\beta]$'s are "specifically those $\alpha$'s that are $\beta$'s"; they are those $\alpha$'s which *must* be $\beta$'s relative to the hypothesis that they are $\alpha$'s (by conditional or relative necessitation). The essential point regarding this special term, one that is central for our present purposes, is that it is such as to validate the inference:

$$\frac{A\alpha\beta}{LA\alpha[\alpha\beta]}$$

[1] For an overview of the current position, together with references to the literature, see Storrs McCall, *Aristotle's Modal Syllogisms* (Amsterdam, 1963) and Nicholas Rescher, "Aristotle's Theory of Modal Syllogisms and Its Interpretation," in *Essays in Philosophical Analysis* (Pittsburgh, 1969), pp. 33–60. For the general background of the Aristotelian syllogistic see Günther Patzig, *Aristotle's Theory of the Syllogism* (Dordrecht, 1968).

Intuitively, if all $\alpha$'s are $\beta$'s, then all $\alpha$'s must be such that they are necessarily $\beta$'s *with a relative necessity*, subject to its being given that they are $\alpha$'s. Correspondingly, we would also have the inference:

$$\frac{I\alpha\beta}{LI\alpha[\alpha\beta]}$$

Intuitively, if some $\alpha$'s are $\beta$'s then some $\alpha$'s are "things which are necessarily $\beta$'s *with a relative necessity*, subject to the condition of their being $\alpha$'s." The bracketed ecthesis-terms are *designed* to make these inferences work.

Such "bracketed terms," as we shall call them provide the materials out of which our interpretation of Aristotle's apodeictic syllogisms will be constructed. Once terms of this type are introduced, it becomes an interesting and significant result that the apodeictic sector of the Aristotelian modal syllogistic follows *in toto* as a natural consequence.

## 2. The Technical Result

Notation and terminology are as in McCall's *Aristotle's Modal Syllogisms* except for the additional primitive use of term-bracketing. Replace McCall's formation rule (i), p. 37, by:

(i′) Any formula $A\alpha\beta$ or $I\alpha\beta$, where $\alpha$ and $\beta$ are terms, is well-formed. Such formulae are called *categorical expressions*.

Here $\alpha$ is to be a term if $\alpha$ is a variable or $\alpha$ is $[\beta\gamma]$ where $\beta$, $\gamma$ are terms. Replace McCall's rule of substitution, p. 37, by:

(i″) Rule of Substitution of terms for variables, where this does not involve identifying terms.[2]

An axiomatization of the assertoric moods *XXX*—and correspondingly of the apodeictic moods *LLL*—in line with the above revisions will be assumed.

In order to extend this basis to include all the apodeictic moods, we adopt the following axiomatic rules with respect to bracketed terms:

*Group 1: Modal Inferences of Type X to L*

I. *C Aab LAa*[*ab*]

II. *C Iab LIa*[*ab*]

*Group 2: Modal Inferences of Type L to L*

III. *C LAab LA*[*ca*]*b*

IV. *C LEab LE*[*ca*]*b*

[2] We shall not attempt to formalize (i″) rigorously, but the intent of (i″) is that (say) *I*[*ab*]*c*, *Ibc*, *I*[*ab*][*cd*], and (even) *I*[*ab*][*ba*] or *I*[*ab*][*bb*] be regarded as substitution instances of *Iab*, but not *Iaa* or *I*[*ab*][*ab*]. For the motivation for restricting McCall's rule of substitution in this way, see R. Z. Parks, "On Formalizing Aristotle's Theory of Modal Syllogisms," *Notre Dame Journal of Formal Logic* vol. 13 (1974), pp. 385–386.

These four rules together with the laws of conversion and of modal conversion yield all the apodeictic moods. To show that all the valid apodeictic moods are desirable on this basis, we shall prove Fitch-style all of those of the first figure:

|  |  |  |  |
|---|---|---|---|
| Barbara *LXL:* | 1 | *LAbc* | hyp |
| | 2 | *Aab* | hyp |
| | 3 | *LAa[ab]* | 2, I |
| | 4 | *LA[ab]c* | 1, III |
| | 5 | *LAac* | 3, 4, Barbara *LLL* |
| Celarent *LXL:* | 1 | *LEbc* | hyp |
| | 2 | *Aab* | hyp |
| | 3 | *LAa[ab]* | 2, I |
| | 4 | *LE[ab]c* | 1, IV |
| | 5 | *LEac* | 3, 4, Celarent *LLL* |
| Darii *LXL:* | 1 | *LAbc* | hyp |
| | 2 | *Iab* | hyp |
| | 3 | *LI[ab]* | 2, II |
| | 4 | *LA[ab]c* | 1, III |
| | 5 | *LIac* | 3, 4, Darii *LLL* |
| Ferio *LXL:* | 1 | *LEbc* | hyp |
| | 2 | *Iab* | hyp |
| | 3 | *LI[ab]* | 2, II |
| | 4 | *LE[ab]c* | 1, IV |
| | 5 | *LOac* | 3, 4, Ferio *LLL* |

It should be remarked that all the derivations follow a perfectly uniform plan, viz., (1) the use of bracketed terms to obtain (using I/II) a modalization from the assertoric minor premiss, in view of which (2) the bracketed term at issue in this minor can be subsumed as a special case under the apodeictic major (using III/IV).[3]

[3] This substantiates the idea of N. Rescher, *op. cit.* (pp. 53–55), that a leading intuition of Aristotle's apodeictic syllogistic is that of a special case falling under a necessary rule:

> "In short, Aristotle espouses the validity of Barbara *LXL* not on grounds of abstract formal logic, but on grounds of *applied* logic, on *epistemological* grounds. What he has in mind is the application of modal syllogisms within the framework of a theory of scientific inference along the lines of his own conceptions. We must recognize that it is Aristotle's concept that in truly scientific reasoning the relationship of major to minor premiss is governed by the proportion:
>
> major premiss: minor premiss:: general rule: special case
>
> When we take note of this line of thought we see why Aristotle taught that the major premiss of a modal syllogism can strengthen the modality of the conclusion above that of the minor premiss. For a rule that is necessarily (say) applicable to all of a group will be necessarily applicable to any sub-group, pretty much regardless of how this sub-group is constituted. On this view, the necessary properties of a genus must necessarily characterize even a con-

The adequacy of any formalization of Aristotle's theory of modal syllogisms depends not only on having the right theorems but also on lacking the wrong ones (and this is where Lukasiewicz fails badly). An important test case is that the theory accept Barbara *LXL* but omit Barbara *XLL*. We are safe on the first count; how do we fare on the second? Let us attempt to prove Barbara *XLL*:

| Barbara *XLL:* | 1 | *Abc* | hyp |
|---|---|---|---|
| | 2 | *LAab* | hyp |
| | | . | |
| | | . | |
| | | . | |
| | n | *LAac* | ? |

Clearly *LAac* is unavailable without the introduction of bracketed terms. Applying rule I to premiss 1 will yield *LAb*[*bc*]. This together with premiss 2 gives us *LAa*[*bc*]—by Barbara *LLL*. But now we are unable to proceed further; we simply cannot infer *LAac* from *LAa*-[*bc*].[4] Since this is in fact our only method of attack, Barbara *XLL* cannot be proven.

The remaining first-figure syllogisms will also be blocked for the type *XLL*. Take Celarent first:

| Celarent X*LL* | 1 | *Ebc* |
|---|---|---|
| | 2 | *LAab* |
| | | . |
| | | . |
| | | . |
| | n | *LEac* |

This is blocked because there is no way of obtaining an L-qualified proposition from an *E*-premiss (or any negative premiss).

Next consider Darii:

| Darii *XLL:* | 1 | *Abc* | |
|---|---|---|---|
| | 2 | *LIab* | |
| | 3 | *LAb*[*bc*] | 1, I |
| | 4 | *LIa*[*bc*] | 2, 3, Darii *LLL* |
| | | . | |
| | | . | |
| | | . | |
| | n | *LIac* | |

tingently differentiated species. If all elms are necessarily deciduous, and all trees in my yard are elms, then all trees in my yard are necessarily deciduous (even though it is not necessary that the trees in my yard be elms). The "special case" subsumption at issue here can be viewed as a mode of application of the *dictum de omni et nullo*." (*Ibid.*, pp. 54–55.)

[4] If all $\alpha$'s are necessarily $\beta$'s-that-in-fact-are-$\gamma$'s, it does not follow that all $\alpha$'s are necessarily $\gamma$'s.

But this inference cannot be accomplished because we cannot infer *LIac* from *LIa*[*bc*].[5]

Finally take Ferio:

Ferio *XLL:*

| | |
|---|---|
| 1 | *Ebc* |
| 2 | *LIab* |
| | . |
| | . |
| | . |
| n | *LOac* |

This inference too is blocked because there is no way of obtaining an L-qualified proposition from an *E*-premiss (or any negative premiss).

It might be noted that the four first figure *XLL* syllogisms are blocked by three principles:

(1) Disallowing the inference of any *L*-qualified proposition from a negative premiss.

(2) Disallowing the inference of *LAac* from *LAa*[*bc*].

(3) Disallowing the inference of *LIac* from *LIa*[*bc*].[6]

These last two principles amount to: *Disallowing the elimination of a bracketed term from an affirmative premiss.*

Thus if appropriate restrictions (of a rather plausible sort) are postulated for inferences involving bracketed terms, none of the apodeictic syllogisms Aristotle regards as illicit will be forthcoming.

If the machinery developed thus far is acceptable from an Aristotelian point of view, we can perhaps explain Aristotle's silence regarding the validity of $LA\alpha\alpha$. If we are to reject $C\ A\alpha\beta\ LA\alpha\beta$ (which one must certainly reject), then given our machinery, we are committed to rejecting $LA\alpha\alpha$.[7] This may be seen as follows:

[5] It deserves note that we cannot, without serious consequences, postulate the nonmodal counterpart of IV, viz., (IV′) *C Eab E*[*ca*]*b* (together with the obvious modal principle that $\vdash C\alpha\beta$ yields $\vdash CL\alpha L\beta$). For IV′ entails *C I*[*ca*]*b Iab*, whose modalized version is *CLI*[*ca*]*b LIab* or equivalently *C LIa*[*bc*] *LIac*. And just this principle must be excluded if Darii *XLL* is to be blocked. It is thus indicated that the assertoric counterparts of III and IV must be rejected, so that these represent specifically apodeictic modes of inference. In summary, by contrast with the acceptable theses I-IV, the following four theses should thus be rejected:

*C LAa*[*bc*] *LAac*
*C LI*[*bc*] *LIac*
*C Aab A*[*ca*]*b*
*C Eab E*[*ca*]*b*

[6] Restrictions (2) and (3) are clearly plausible. If all or some $\alpha$'s are $\gamma$'s, that does not mean they must necessarily be members of the $\gamma$-species of $\beta$'s.

[7] In consequence of this rejection, it would no longer be necessary to introduce the above-mentioned restriction on McCall's rule of substitution.

| | | |
|---|---|---|
| 1 | $Aab$ | hypothesis |
| 2 | $LAa[ab]$ | 1, I |
| 3 | $LAbb$ | by the thesis at issue |
| 4 | $LA[ab]b$ | 3, III |
| 5 | $LAab$ | 2, 4, Barbara $LLL$ |

This serves to motivate omission of $LA\alpha\alpha$. We can only explain the lack of an explicit rejection by saying that if one must reject $LAaa$, one might well prefer doing so quietly. (Though if one is enough of an essentialist, it would seem not incongruous to take the view that among all the $\alpha$'s some should be $\alpha$'s of necessity but others merely by accident, so that $LA\alpha\alpha$ would not be acceptable.)[8] Although the Aristotelian modal syllogistic must reject the thesis $LA\alpha\alpha$, the cognate thesis $LI\alpha[\alpha\alpha]$ is readily demonstrable:

| | | |
|---|---|---|
| 1 | $Aaa$ | thesis |
| 2 | $LIAa[aa]$ | 1 by ecthesis |

Actually, although a strict proof does not seem available, it would appear that $LA[\alpha\alpha][\alpha\alpha]$—and indeed even $LA[\alpha\beta][\alpha\beta]$—could well be viewed as acceptable theses.

It is worthwhile to point out that the system suggested here is consistent. We define a function h inductively as follows: (a) if $\alpha$ is a variable, $h(\alpha) = \alpha$, (b) $h([\alpha\beta]) = h(\beta)$, (c) $h(R\alpha B) = Ah(\alpha)h(\beta)$, (d) $h(I\alpha\beta) = Ih(\alpha)h(\beta)$, (e) $h(N\alpha) = Nh(\alpha)$, (f) $h(L\alpha) = Lh(\alpha)$ and (g) $h(C\alpha\beta) = Ch(\alpha)h(\beta)$. Clearly, if $\alpha$ is a theorem, $h(\alpha)$ is a theorem of the assertoric theory of the syllogism. So, our system is consistent if the assertoric *theory* is. But the latter is consistent.[9]

## 3. Scientific Demonstration

The deductions given above for the valid first-figure syllogisms of type $XLX$ enable us to clear up one of the puzzles of Aristotle-interpretation: the discord between the theory of apodeictic inference presented in *Prior Analytics* and that presented in *Posterior Analytics*.

Aristotle's theory of modal syllogisms has its developmental and conceptual roots in Aristotle's concept of demonstration. To understand the role and nature of the Aristotelian theory of modal syllogisms generally, and of apodeictic syllogisms in particular, we are thus well advised to look to his concept of scientific reasoning in *Anal. Post.*, for it is here that we learn clearly and explicitly how Aristotle

[8] Previous attempts to formalize Aristotle's modal syllogic (specifically those of Lukasiewicz and McCall) also explicitly reject $LA\alpha\alpha$. See Jan Lukasiewicz, *Aristotle's Syllogistic*, 2nd edition (Oxford, 1957), p. 190, and Storrs McCall, *op. cit.*, p. 50.

[9] See, for example, J. C. Shepherdson's "On The Interpretation of Aristotelian Syllogistic," *Journal of Symbolic Logic*, vol. 21 (1956), pp. 137–147.

conceives of the job and function of apodeictic reasoning. The following passages (quoted from the Oxford translation of Aristotle's works) should prove sufficient for our present purposes.

> By demonstration I mean a syllogism productive of scientific knowledge, a syllogism, that is, the grasp of which is *eo ipso* such knowledge. Assuming then that my thesis as to the nature of scientific knowing is correct, the premisses of demonstrated knowledge must be true, primary, immediate, better known than and prior to the conclusion, which is further related to them as effect to cause. Unless these conditions are satisfied, the basic truths [of a demonstration] will not be "appropriate" to the conclusion. (71b17–24.)
>
> Since the object of pure scientific knowledge cannot be other than it is, the truth obtained by demonstrative knowledge will be necessary. And since demonstrative knowledge is only present when we have a demonstration, it follows that demonstration is an inference from necessary premisses. (73a20–24.)
>
> Demonstrative knowledge must rest on necessary basic truths; for the object of scientific knowledge [i.e., that which is known by demonstration] cannot be other than it is. Now attributes attaching essentially to their subjects attach necessarily to them. . . . It follows from this that the premisses of the demonstrative syllogism must be connections essential in the sense explained. . . . We must either state the case thus, or else premise that the conclusion of demonstration is necessary and that a demonstrated conclusion cannot be other than it is, and then infer that the conclusion must be developed from necessary premisses. For though you may reason from true premisses without demonstrating, yet if your premisses are necessary you will assuredly demonstrate—in such necessity you have at once a distinctive character of demonstration. That demonstration proceeds from necessary premisses is also indicated by the fact that the objection we raise against a professed demonstration is that a premiss of it is not a necessary truth . . . A further proof that the conclusion must be developed from necessary premisses is as follows. Where demonstration is possible, one who can give no account which includes the cause has no scientific knowledge. If, then, we suppose a syllogism in which, though A necessarily inheres in C, yet B, the middle term of the demonstration, is not necessarily connected with A and C, then the man who argues thus has no reasoned knowledge of the conclusion, since this conclusion does not owe its necessity to the middle term; for though the conclusion is necessary, the mediating link is a contingent fact. . . . To sum up, then: demonstrative knowledge must be knowledge of a necessary nexus, and therefore must clearly be obtained through a necessary middle term; otherwise its possessor will know neither the cause nor the fact that this conclusion is a necessary connection. (74b5–75a15.)

Note that in these discussions in *Anal. Post.* Aristotle adopts the view that a syllogistic result is necessary only when *both* premisses are necessary. This, of course, is a view he does not espouse in *Anal. Pr.*, holding there that it suffices (with first-figure syllogisms) for the major premiss alone to be necessary. No statement could be more sharply explicit than the following: "All demonstrative reasoning proceeds from necessary or general premisses, the conclusion being necessary if the premisses are necessary and general if the premisses are general." (*Anal. Post.*, 87b23–25.)

I am not aware that any Aristotelian student, modern or ancient, has noted, let alone assessed, the significance of this variation of Aristotle's conception of necessary inference as between the *Prior* and the *Posterior Analytics*.

From our present perspective, however, it becomes easy to see how the discrepancy at issue can be resolved. For in all cases, the reasoning at issue in establishing the valid first-figure *LXL* syllogism proceeded by (1) making an *immediate* inference on the assertoric premiss to transform it into a necessary premiss (viz., one with a bracketed term), and from there (2) deriving the desired apodeictic conclusion from an *LLL*-syllogism. In short, *the only syllogistic inference operative in the course of demonstration* is indeed of the all-premisses-necessary type. In effect, the ultimate justification of an apodeictic inference with an assertoric premiss lies in our capacity to subsume the argument at issue within the framework of a "scientific" syllogism of strictly necessitarian proportions.

The fundamental fact is thus that the "scientific syllogism" of our demonstrative reasoning *must* have necessary premisses. To be sure, apodeictic conclusions can also be established in certain suitable cases where assertoric premisses occur. But even here the basis of necessity is in the final analysis mediated by and realized through a genuine scientific demonstration—one that is *altogether* necessitarian. Once the matter is seen in this light we obtain a plausible reconciliation between the seemingly conflicting Aristotelian views that a scientifically demonstrative inference must proceed from necessary premisses, and yet that an apodeictic conclusion can be obtained when an assertoric premiss is present.[10]

## 4. Ecthesis

Aristotle does not give proofs for Baroco *LLL* and Bocardo *LLL* but merely outlines how they are to proceed (*An. pr.*, i. 8, 30a6). Both are to be proven by ecthesis.

[10] Note that the modally mixed inference at issue must still be classed as "perfect" in a syllogistic context: it must be taken to be one of the fundamental postulates of the syllogistic system rather than viewed as a derivable consequence. It is nonderivative from a logical point of view because the "immediate inference" that renders it derivable is metaphysical, not logical. And it remains syllogistically "perfect" because there is no way of twisting or turning so as to validate it by syllogistic means.

We propose to construe this process—which Aristotle leaves somewhat mysterious—along the following lines:

(1) Nonmodal ecthesis

$$\frac{I\alpha\beta}{(\exists\gamma)[KA\gamma\alpha A\gamma\beta]} \qquad \frac{O\alpha\beta}{(\exists\gamma)(KA\gamma\alpha E\gamma\beta)}$$

(2) Modal ecthesis

$$\frac{LI\alpha\beta}{(\exists\gamma)(KLA[\alpha\gamma]\alpha LA[\alpha\gamma]\beta)} \qquad \frac{LO\alpha\beta}{(\exists\gamma)(KLA[\alpha\gamma]\alpha LE[\alpha\gamma]\beta)}$$

Ecthesis, thus conceived, is a process for inferring universal propositions from particulars.[11] Its central feature in the modal case is its recourse to bracketed terms as introduced above. (It might be noted that the inferences in (1) and (2) are to be reversible into corresponding inverse forms. Thus our construal of non-modal ecthesis coincides with that of Patzig.[12] Aristotle's observations at *An. pr.*, i. 6, $28^a$ 22–26, are simply a *statement* of the inverse form of the affirmative case of nonmodal ecthesis, rather than representing—as W. D. Ross complains—an attempt at "merely proving one third-figure syllogism by means of another which is no more obviously valid."[13]

Let us examine the argument for Baroco *LLL* as Ross[14] presents it. According to Ross (p. 317) the proof goes as follows: assume that all *B* is necessarily *A* and that some C is necessarily not *A*. Take some species of *C* (say *D*) which is necessarily not *A*. Then all *B* is necessarily *A*, all *D* is necessarily not *A*, therefore all *D* is necessarily not *B* (by Camestres *LLL*). Therefore some *C* is necessarily not *B*. The reasoning may be formulated as follows:

| | | |
|---|---|---|
| 1 | $LAba$ | hyp |
| 2 | $LOca$ | hyp |
| 3 | $(\exists d)LE[cd]a$ | ecthesis on 2 |
| 4 | $(\exists d)LE[cd]b$ | 1, 3, Camestres *LLL* |
| 5 | $LOcb$ | 4, inverse ecthesis |

Next, consider the argument for Bocardo *LLL*. Ross (*ibid.*) construes the argument as follows: assume that some *C* is necessarily not *A* and that all *C* is necessarily *B*. Take a species of *C* (say *D*) which is necessarily not *A*. Then all *D* is necessarily not *A*, all *D* is necessarily

[11] The inverse inferences (closely akin to Darapti and Felapton), are, of course also valid, so that we are, in effect, dealing with equivalences.

[12] Cf. Günther Patzig, *Aristotle's Theory of the Syllogism* (Dordrecht, 1968), pp. 156–168. In support of his interpretation of nonmodal ecthesis, Patzig cites *Anal. Pr.*, i.28, 43b43–44a2 and 44a9–11, which appears to be a statement of the equivalence of the premisses and their respective conclusions in (1).

[13] W. D. Ross, *Aristotle's Prior and Posterior Analytics* (Oxford, 1949), p. 32.

[14] W. D. Ross, *ibid.*

*B*, therefore some *B* is necessarily not *A* (by Felapton *LLL*). The reasoning also is readily formalized as follows:

| | | |
|---|---|---|
| 1 | $LOca$ | hyp |
| 2 | $LAcb$ | hyp |
| 3 | $(\exists d)LE[cd]a$ | ecthesis on 1 |
| 4 | $(\exists d)LE[dc]a$ | 3 (supposing $E[\alpha\beta]\gamma$ yields $E[\beta\alpha]\gamma$) |
| 5 | $(\forall d)LA[dc]b$ | 2, III |
| 6 | $LOba$ | 4, 5, Felapton *LLL* |

The use of bracketed terms to explicate ecthesis along the lines outlined above thus provides a simple way to systematize the Aristotelian justification of certain apodeictic syllogisms.

One further point. Consider the argumentation in detail. Suppose a concrete case of the premisses:

Major (Law): All *b* is necessarily *c* ($LAbc$)
Minor (Special Case): All *a* is *b* ($Aab$)

In developing the line of derivational argument, we proceed as follows. We use the considerations adduced above to establish *in abstracto* the *existence* of a term $\xi$ (viz., [*ab*]) such that

$$LAa\xi$$
$$LA\xi c$$

and now proceed by strictly "scientific syllogism" to demonstrate our conclusion

$$LAac$$

But, of course, quantifiers are not to be used in Aristotle's syllogistic. To carry out the demonstrative argument we need to *find* the specific middle term whose existence the abstract considerations guarantee. Just this consideration serves to provide a justificatory rationale for Aristotle's construction of the provision of scientific explanations in terms of the search for a middle term and for his assimilation of this search for the middle to the problem of identifying causes.

## 5. Conclusion

The use of bracketed terms in connection with modal and ecthesis-involving reasonings is analogous in one significant respect: In both cases their introduction allows us "to do the impossible" in Aristotelian logic—albeit in a perfectly legitimate way. In the one case we move from an assertoric to an apodeictic proposition:

$$\frac{A\alpha\beta}{LA\alpha[\alpha\beta]}$$

In the other case we move from a particular to a universal proposition:

$$\frac{LI\alpha\beta}{(\exists\gamma)LA[\alpha\gamma]\beta}$$

In both cases the bracketing operator enables us to "select" from among all the $\alpha$'s those which—given that a certain relationship holds between the $\alpha$'s and $\beta$'s—bear a yet more stringent relation to the $\beta$'s than the $\alpha$'s in general do.

The just-indicated argument paradigm

$$\frac{LI\alpha\beta}{(\exists\gamma)LA[\alpha\gamma]\beta}$$

deserves further comment. It is crucial that the particularized relation the premiss lays down between $\alpha$ and $\beta$ (their *I*-linkage) is necessary, otherwise the conclusion would clearly not be forthcoming. Thus *perception*—which can establish particular linkages *de facto* but not necessarily—cannot provide scientific knowledge.[15] Chance conjunctions in general cannot in the very nature of things be subject to demonstrations of necessity.[16]

That nonmodal ecthesis is a logically warranted (indeed virtually trivial) process can be seen along the following lines:

1. Assume by way of hypothesis that: Some *a* is b
2. Let $x_1, x_2, \ldots$ be specifically those *a*'s that are *b*'s and let us designate the group of these $x_i$, the "*a*'s at issue," as $\xi$.
3. Then all these $\xi$'s are *a*'s (by definition of $\xi$) and moreover all $\xi$'s are *b*'s, and conversely (for the same reason).

Thus between the "*a*'s at issue," viz., $x_1, x_2, \ldots$, and *b* we have inserted a "middle term" ($\xi$) in such a way that (1) All the "*a*'s at issue" are $\xi$'s (and conversely) and (2) All $\xi$'s are *b*'s. No doubt here, in the assertoric (nonmodal) case, we have done this insertion in a logically trivial way.

But in the modal case, when *Some a is necessarily b* the issue of inserting an intermediate $\xi$ such that both *All the a's at issue are* $\xi$ and *All* $\xi$ *is necessarily b* is not trivial at all. For whereas the motivation of the first of the two inferences under consideration is essentially a matter of pure logic that of the second is at bottom not logical, but metaphysical. If some $\alpha$'s are necessarily $\beta$'s, then—so the inference has it—there must be some $\alpha$-delimitative species, the $[\alpha\gamma]$'s, *all* of which are necessarily $\beta$'s. If some metals are necessarily magnet-attracted then there must be a type of metal (e.g., iron) all of which

[15] Cf. *Anal. Post.*, I, 31.
[16] *Ibid.*

is necessarily magnet-attracted. The governing intuition here operative lies deep in the philosophy of nature: Whenever $\alpha$'s are such that some of them must be $\beta$'s, then this fact is capable of *rationalization*, i.e., there must in principle be a *natural kind* of $\alpha$'s that are necessarily (essentially, lawfully) $\beta$'s.

A precursor version of the principle of causality is at work here: If some "men exposed to a certain virus" are in (the naturally necessitated course of things) "men who contract a certain disease," but some are not, then there must be some *characteristic* present within the former group in virtue of which those of its members exhibiting this characteristic *must all* contract the disease if exposed to it. To explain that some $\alpha$'s have to be $\beta$'s we must find a naturally constituted species of the $\alpha$'s all the members of which are necessarily $\beta$'s.[17] Thus given "Some $\alpha$'s are of necessity $\beta$'s," it follows from the requisites of explanatory rationalization that for some species $\gamma$ of the $\alpha$'s we have "All $\gamma$'s are necessarily $\beta$'s." We come here to what is essentially not a principle of logic but a metaphysical principle of rationalization. At this precise juncture, the logic of the matter is applied rather than pure—fusing with the theory of scientific explanation presented in *Posterior Analytics*.

From this standpoint, then, the principle of modal ecthesis

$$\frac{LI\alpha\beta}{(\exists\gamma)KLA[\alpha\gamma]LA[\alpha\gamma]\beta}$$

is based upon metaphysical rather than strictly logical considerations. This principle underwrites the equivalence:

$$\text{LI}\alpha\beta \text{ if and only if } (\exists\gamma)\text{LA}[\alpha\gamma]\beta$$

This, in effect, is a "generalization principle for necessary connection." It stipulates that whenever a necessary connection exists between two particular groups $\alpha$ and $\beta$ the matter cannot rest there. There must be—somehow, no matter how well concealed—a *universal* necessary relationship from which this particular case derives and in which it inheres. There can be no particular necessity as such: necessity, whenever encountered, is always a specific instance of a *universal* necessity. It is thus easy to see the basis for Aristotle's policy (in *Posterior Analytics* and elsewhere) of assimilating necessity to universality. This perspective highlights Aristotle's fundamental position that science, since it deals with the necessary, cannot but deal with the

[17] The idea is closely analogous with the "generalization principle" in modern ethics, i.e., the thesis that if some certain men are obligated (or entitled) to do something, then this must be so because they belong to a group *all* of whose members are obligated (or entitled) to do so.

universal as well. The irreducibly particular—the accidental—lies wholly outside the sphere of scientific rationalization.

Insofar as this view of the matter has merit, it stresses the conclusion that the fundamental motivation of Aristotle's modal syllogistic is heavily indebted to metaphysical rather than strictly logical considerations. Be this as it may, it is, in any case, significant that by introducing such an ecthesis-related specification of terms, the apodeictic sector of Aristotle's modal syllogistic is capable of complete and straightforward systematization.[18]

[18] This essay is an expanded reworking of a paper of the same title written in collaboration with Zane Parks and first published in *The Review of Metaphysics*, vol. 24 (1971), pp. 678–689.

# II.

# The Theory of Modal Syllogistic in Medieval Arabic Philosophy

## 1. Introduction

IN the wake of A. N. Prior's book on *Time and Modality*[1] an active interest has sprung up among logicians and philosophers in the logical theory of chronological propositions generally, and particularly in the relationships that obtain between such propositions and modal concepts. This phenomenon is not surprising, because the issue is one that ramifies widely into various topics of logico:-philoso phical interest: the theory of tensed discourse, the problem of determinism, and the puzzle of future contingency, among others. The modern discussions have gone forward wholly oblivious to the fact that medieval Arabic logicians had given extensive attention to the development of a theory of temporal modalities, and had developed an extensive and subtle machinery for dealing with problems in this area. The aim of the present discussion is one of "intellectual archeology" —to present the Arabic contributions to this branch of logic in such a way that their linkage with ideas and concepts of present-day interest can be assessed and appreciated.

The logic of modality was a relative latecomer to the domain of Arabic logic. Only after the time of Abū Bishr Mattā ibn Yūnus (ca. 870–940), translator of *Posterior Analytics*, did the study of Aristotle's modal syllogistic come to be taken up.[2] In its wake, interest in other Greek ideas regarding modality sprung up, and in the 10th century there was a lively polemic in the School of Baghdad—whose principal figure was al-Fārābī (ca. 863–950)—against Galen's views on modality, especially his rejection of the modality of possibility.[3] The mainstream of Arabic logic which stayed within the tradition of the School of Baghdad, culminating in Averroes (1126–1198), always

[1] Oxford, The Clarendon Press, 1957.

[2] "Al-Fārābī on Logical Tradition" in N. Rescher, *Studies in the History of Arabic Logic* (Pittsburgh, 1963), pp. 21–27.

[3] See N. Rescher, *The Development of Arabic Logic* (Pittsburgh, 1964), p, 43.

remained closely faithful to Aristotelian views.[4] On the other hand, the influence of Galen, and especially—perhaps through his mediation—of the Stocis, made significant headway in that part of the Arabic logical tradition whose foundation-head was Avicenna.[5] But while Avicenna gave the theory of temporal modalities its principal development and its main impetus, he cannot be viewed as its inventor. For there can be little doubt that the theory of temporal modalities with which we shall be dealing in the present essay finds its origins in Greek, and above all in Stoic logic. We shall, however, postpone any detailed consideration of the cognate conceptions of Megarian and Stoic thought until later, after our examination of the Arabic materials has been completed.

The theory of temporalized modalities was developed by the Arabs within an Aristotelian setting, against the backdrop of a concept of demonstrative science insisting upon explanations in terms of premisses that are necessary in the sense of comprising constantly operative causes referring to what *always* happens.[6] Consider, for the sake of illustration, Avicenna's contention:

> The mind is not all repelled by the statement, "when Zayd moved his hand, the key moved," or "Zayd moved his hand, then the key moved." The mind is repelled, however, by the statement, "when the key moved, Zayd moved his hand," even though it is [rightly] said, "when the key moved, *we knew that* Zayd moved his hand." The mind, despite the temporal coexistence of the two movements, assigns a (causal) priority for one, a posteriority for the other. For it is not the existence of the second movement that causes the existence of the first; it is the first movement that causes the second.[7]

Its rooting in the fertile soil of the concepts of necessity, causality and scientific explanation provided for the Arabs the impetus to preoccupation with the logical theory of temporally modalized propositions.[8]

[4] See, for example, "Averroes' *Quaesitum* on Assertoric (Absolute) Propositions" in N. Rescher, *Studies in the History of Arabic Logic* (*op. cit.*), pp. 91–105. The extent to which the Spanish Muslim logicians were faithful to Aristotle is exemplified by the treatment of modal syllogistic by Abū 'l-Ṣalt (1068–1134). See "Abū–'l-Ṣalt of Denia on Modal Syllogisms," *ibid.*, pp. 87–90.

[5] See, for example, "Avicenna on the Logic of 'Conditional' Propositions," *ibid.*, pp. 76–86. On the general phenomenon see N. Rescher, *The Development of Arabic Logic*, *op. cit.*, pp. 50 ff., and also *idem*, *Galen the Syllogism* (Pittsburgh, 1966), pp. 4–8.

[6] Cf. M. E. Marmura, "Ghazali and Demonstrative Science," *Journal of the History of Philosophy*, vol. 3 (1965), pp. 183–204.

[7] *Ibn Sīnā*, *Al-Shifā'*: *al-Ilāhīyāt* (*Metaphysics*) (ed. C. G. Anawati, S. Dunya, and S. Zayd, revised by M. Madkur), 2 vols. (Cairo, 1960), vol. I, p. 165.

[8] The earliest exposition of the theory of temporal modal syllogistic as expounded by the medieval Arabic philosophers is N. Rescher, *Temporal Modalities in*

The critical aspect of a theory of temporal modalities that revolves about what *always* (or *mostly* or *sometimes* or *never*) happens when some condition is realized is that it is totally *actualistic* in its orientation. Its horizons are limited to the actual world and what happens in it, and in this sense it is predicated on the Aristotelian dictum that actuality is prior to possibility (and to necessity as well). The whole spectrum of modal ideas can be articulated without reference to the conception of unrealized possibilities populating other possible worlds distinct from this one (ideas that never really obtained a secure foothold in Greek philosophy, nor in its Arabic successor).

## 2. The Textual Basis

The present essay seeks to present in systematic form the theory of modal temporal syllogistic as it is presented in tracts by two authors, al-Qazwīnī and al-Shirwānī. It goes perhaps without saying, that although we base our discussion in the main upon the treatises of these two authors, the exposition will draw upon a wider range of Arabic logicians, including such more prominent figures as Avicenna and Averroes. In particular, we shall see that al-Qazwīnī al-Kātibī follows and draws upon Avicenna. Details of this reliance will be obvious at many points to anyone who compares al-Qazwīnī's treatment with the corresponding treatment of Avicenna's *Kitāb al-ishārāt wa-'l-tanbīhāt*, which is fortunately accessible to European scholars in an excellent French version: A. M. Goichon (tr.), *Ibn Sīnā: Livre des Directives et Remarques* (Beyrouth and Paris, 1951). Al-Qazwīnī's dependence upon Avicenna is strikingly evidenced by the close parallelism in organization, mode of treatment, and substance between the two treaties.

One principal basis for our discussion is "The Sun Epistle" *Al-Risālah al-shamsiyyah* of the thirteenth-century Persian philosopher-scientist al-Qazwīnī al-Kātibī[9] (ca. 1220–1276 or 1292). Not only is this work one of the few that treats of our problem in significant detail, but it is one of the very few Arabic logic treatises to have been

---

*Arabic Logic* (Dordrecht, 1967; *Foundations of Language* Supplementary series, No. 2). Some supplements to this material were presented in N. Rescher, *Studies in Arabic Philosophy* (Pittsburgh, 1968), chapters VII–VIII. Substantial improvements in the description of the theory were made in N. Rescher and A. vander Nat, "New Light on the Arabic Theory of Temporal Modal Syllogistic," in G. Hourani (ed.), *Essays in Islamic Philosophy and Science* (Albany, 1973). The present discussion is drawn in substantial part from this last-named work.

[9] For this Arabic logician—as well as all others to be mentioned here—see the biobibliographical register in N. Rescher, *The Developments of Arabic Logic* (*op. cit.*); see pp. 203–204.

put into a European language, as follows: Aloys Sprenger, *Dictionary of the Technical Terms Used in the Sciences of the Musulmans*, Part 2 (Calcutta, 1862). Appendix I (1862) on "The Logic of the Arabians" gives an Arabic text edition and a (somewhat incomplete) English translation of our treatise. This translation has a serious shortcoming from the standpoint of our present concerns. As the translator explains, certain parts of the treatise (to wit, the following sections, Engl. 68–70 = Ar. 66–68; Engl. 72–74 = Ar. 70–72; Engl. 84–86 = Ar. 81–84) "are omitted in the translation because they contain details on modals which are of no interest. The last-named four paragraphs [dealing with modal syllogisms] are also omitted in most Arabic texts books on Logic, and not studied in Mohammedan Schools" (p. 25, n. 39). (I have presented an English translation of this material in another publication to which the present chapter is substantially indebted: *Temporal Modalities in Arabic Logic* [Dordrecht, 1966; Supplementary Series of *Foundations of Language*].) No matter how difficult or boring such considerations may have proved for the Muslim schoolmaster, it is of the greatest relevance for our interests.

Arabic manuscript codex OR12405 of the British Museum contains a logical treatise entitled *Sharḥ Al-takmīl fī 'l-manṭiq*, whose author is one Maḥammad ibn Fayḍ Allah ibn Muḥammad Amīn al-Shirwānī.[10] Nothing further is independently known about him,[11] apart from what can be gleaned from this manuscript itself. The codex contains two treatises by this scholar, written in the author's own hand, in a somewhat cramped naskhī of 23 lines per folio. In addition to the text at issue (in folios 72–104), it contains also (in folios 1–70) his commentary on the well-known tract *Al-Hashiyyah* (or *Al-Risālah*) *al-sughrā fī 'l-manṭiq* of 'Alī ibn Muḥammad al Jurjāni al-Sayyid al-Sharīf (1340–1413).[12] Al-Shirwānī is thus a late medieval Persian scholar of presumably the early 15th century who must be considered obscure in view of his near-total absence from the manuscript tradition. One item of biographical information which can be gleaned

[10] During the academic year 1967–1968, I had the opportunity of spending a sabbatical term in England with the support of a grant-in-aid from the American Philosophical Society to examine the Arabic logical manuscripts of several libraries, the British Museum in particular. This occasioned contact with the manuscript now at issue, and the assistance of the American Philosophical Society is herewith gratefully acknowledged. At this point I should also like to thank Mr. Zakaria Bashier for his help in translating and interpreting several passages of Shirwānī's text.

[11] He is nowhere mentioned in Brockelmann's *Geschichte der Arabischen Litteratur*.

[12] For this logician see N. Rescher, *The Development of Arabic Logic* (Pittsburgh, 1964), pp. 222–223.

from our text is that the author is the great-grandson of al-Ṣadr al-Shirwānī Muḥammad Ṣadiq ibn Fayḍ Allah ibn Muḥammad Amīn, also otherwise unknown. Al-Shirwānī's treatise is of substantial interest because it enables us to confirm and extend in significant ways our information regarding the Arabic theory of temporal modal syllogistic as available from other sources.[13] What is surprising is not that some minor error should from time to time arise, but that—given the total absence of formalization and even abbreviative devices—these errors are so rare.

## 3. Basic Elements of the Arabic Theory of Temporal Modalities: The Simple Modes

The theory of temporal modal syllogisms as presented in Arabic logical texts further qualifies the relation that the predicate bears to the subject in the four basic categorical propositions **A** (All *A* is *B*"), **E** ("No *A* is *B*"), **I** "Some *A* is *B*"), and **O** ("Some *A* is not *B*") in certain characteristic ways. For *simple* modal propositions two qualifications are added:

(1) a *modality* of one of the following four types

- i. ($\Box$): of necessity
- ii. ($\Diamond$): by a possibility
- iii. ($\forall$): in perpetuity
- iv. ($\exists$): in (some) actuality, or, sometimes

(2) a *temporality* qualifying the modality of one of the following four types

- i. ($\mathscr{E}$): when the subject at issue exists; that is, during times of the existence of the subject.
- ii. ($\mathscr{C}$): when the subject at issue exists and meets a certain condition as specified by the subject term of the proposition; that is, during times of the existence of the subject when it meets the condition stipulated by the subject term.
- iii. ($\mathscr{T}$): when the subject at issue exists during a definite, specifiable time; that is, during a certain *specified* and determinate period of the existence of the subject (e.g., its youth).

[13] N. Rescher, *Temporal Modalities in Arabic Logic* (Dordrecht, 1966; *Foundations of Language*, Supplementary Series, no. 2) is the basic publication, although the data presented there are extended and amplified in chapters VII–VIII of *idem*, *Studies in Arabic Philosophy* (Pittsburgh, 1967). The materials with which the present paper deals makes it possible not only to extend but also in important ways to correct the presentation of the theory given in these earlier discussions.

iv. ($\mathscr{S}$): when the subject at issue exists during some indefinite, unspecifiable time; that is, during some *unspecified* and indeterminate period of the existence of the subject.

Note that the temporalities $\mathscr{T}$ and $\mathscr{S}$ being inherently time-restricted, do not allow of further qualification by the specifically temporal modalities $\forall$ and $\exists$.

In "order of strength" the four modalities are arranged as $\Box$, $\forall$, $\exists$, $\Diamond$, and may be termed necessity, perpetuity, actuality, and possibility, respectively. The order of strength of the temporalities $\mathscr{E}$, $\mathscr{C}$, $\mathscr{T}$, $\mathscr{S}$ depends on their combination with modality. (Concerning these relative strengths see section VI below.) The four temporalities may be called the existential, the conditional, the temporal, and the spread temporality, respectively. Examples of categorical propositions displaying these temporalities are:

$\mathscr{E}$: All men are animals, as long as they exist.
$\mathscr{C}$: All writers move their fingers, as long as they write.
$\mathscr{T}$: All moons are eclipsed at the time when the earth is between it and the sun.
$\mathscr{S}$: All men breathe at some times.

In these examples modality has, for simplicity, been put aside.

The (simple) modal propositions arrived at by the full-scale use of this machinery are as shown in Table I.

Table I

STANDARD EXAMPLES OF SIMPLE MODES

| | |
|---|---|
| ($\Box\mathscr{E}$): | All men are rational of necessity (as long as they exist).[14] |
| ($\Box\mathscr{C}$): | All writers move their fingers of necessity as long as they write. |
| ($\Box\mathscr{T}$): | The moon is eclipsed of necessity at the time when the earth is between it and the sun. |
| ($\Box\mathscr{S}$): | All men breathe of necessity at some times. |
| ($\forall\mathscr{E}$): | All men are rational perpetually (as long as they exist). |
| ($\forall\mathscr{C}$): | All writers move as long as they write. |
| ($\exists\mathscr{C}$): | All writers move while they are writing. |
| ($\mathscr{T}$): | All writers move at the time they are writing. |
| ($\mathscr{S}$): | All men breathe at certain times. |
| ($\exists\mathscr{E}$): | All men breathe (at some times). |
| ($\Diamond\mathscr{C}$): | All writers move with a possibility while they are writing. |
| ($\Diamond\mathscr{T}$): | The moon is eclipsed with a possibility at the time when the earth is between it and the sun. |
| ($\Diamond\mathscr{S}$): | All men breathe with a possibility at all times. |
| ($\Diamond\mathscr{E}$): | All writers move with a possibility (at some time). |

[14] The existence condition is usually unstated.

By ringing the changes on the two factors of modality and temporality, fourteen theoretical combinations arise. Six of these, ($\Box\mathscr{E}$), ($\Box\mathscr{C}$), ($\forall\mathscr{E}$), ($\forall\mathscr{C}$), ($\exists\mathscr{E}$), and ($\Diamond\mathscr{E}$) are explicitly listed and discussed by al-Qazwīnī, and he refers also to others, namely ($\Box\mathscr{T}$), ($\Box\mathscr{S}$), ($\exists\mathscr{C}$), and ($\Diamond\mathscr{C}$), though not giving them an explicit place in his inventory.[15] Al-Shirwānī explicitly recognizes all fourteen, and his presentation of them is summarized in Table II. The more detailed analysis of these modal propositions is deferred until section 6 below. In particular, the relationship of immediate inference in terms of the relative strength of these modal propositions (as summarized in Table VII below) serve to throw light on the logical relationships among these modal propositions.

In our earlier publication,[16] the analysis of the Arabic temporal modalities was based on the *Risālah al-shamsiyyah*, the *Sun Epistle* of

Table II

SIMPLE MODES IN SHIRWANI

| *Type* | *Name* |
|---|---|
| I. MODES OF NECESSITY | |
| $\Box\mathscr{E}$ | absolute necessary |
| $\Box\mathscr{C}$ | general conditional |
| $\Box\mathscr{T}$ | absolute temporal (#) |
| $\Box\mathscr{S}$ | absolute spread (#) |
| II. MODES OF PERPETUALITY | |
| $\forall\mathscr{E}$ | absolute perpetual |
| $\forall\mathscr{C}$ | general conventional |
| III. MODES OF ACTUALITY | |
| $\exists\mathscr{E}$ | general absolute |
| $\exists\mathscr{C}$ | absolute temporary or absolute continuing (#) |
| $\mathscr{T}$ | temporal absolute (*) |
| $\mathscr{S}$ | spread absolute (*) |
| IV. MODES OF POSSIBILITY | |
| $\Diamond\mathscr{E}$ | general possible |
| $\Diamond\mathscr{C}$ | possible continuing (#) |
| $\Diamond\mathscr{T}$ | temporal possible (*) |
| $\Diamond\mathscr{S}$ | spread possible or perpetual possible (*) |

NOTE: An asterisk (*) marks those modes missing in Qazwīnī and (#) marks those which are recognized by Qazwīnī but not listed or discussed by him.

[15] It would seem that the six modes are considered by al-Qazwīnī to be the *standard* modes—modes, as he says, "into which it is usual to inquire."

[16] N. Rescher, *Temporal Modalities in Arabic Logic*, *op. cit.*

al-Qazwīnī al-Katibī.[17] Al-Qazwīnī's treatment differs from that of al-Shirwānī in that in Qazwīnī the temporalities ($\mathscr{S}$) and ($\mathscr{T}$) never occur with simple but *only* with compound propositions, and also in that Qazwīnī's analysis does not include the mode ($\exists\mathscr{C}$). Presumably, Qazwīnī assimilated these simple modes under the temporality condition ($\mathscr{E}$). (So with Ibn al-Assāl [ca. 1190–1250], who appears to assimilate the weak modes ($\exists\mathscr{C}$) and ($\Diamond\mathscr{C}$) and ($\Diamond\mathscr{T}$) with ($\exists\mathscr{E}$) and ($\Diamond\mathscr{E}$), respectively.[18]) Apart from this difference, there is complete agreement between Qazwīnī and Shirwānī regarding the nature and nomenclature of simple modes. Shirwānī's treatment in effect extends Qazwīnī's on the side of temporal conditionalization.[19]

## 4. Negation and Conversion for Simple Modalities

The rule of negation for simple modal propositions is as follows. Let the initial proposition to be negated take the form

$$(\text{modality/temporality})P$$

Then its contradictory takes the form

$$(\text{o-modality/temporality})\sim P$$

Here the o-modality is the *modal opposite* of the initial modality (formed by interchanging $\Box$ and $\Diamond$ on the one hand and $\forall$ and $\exists$ on the other). Moreover, the initial categorical proposition $P$ is replaced by its *contradictory* $\sim P$, and the temporality remains unchanged.

It must be noted at this juncture that in analyzing the modal propositions of Shirwānī we are dealing with *modes of predication* (modality *de re*) rather than with strictly propositional modes (modality *de dicto*): the issue is one of qualifying the relation of the predicate to the subject rather than qualifying an entire categorical proposition. For example, the modal proposition

(1) ($\forall\mathscr{E}$) (All men are animals)

is to be understood as

(2) All men are always animals

[17] For this writer see N. Rescher, *The Development of Arabic Logic* (*op. cit.*), pp. 203–204. Appendix I of Aloys Springer's *Dictionary of the Technical Terms Used in the Sciences of the Musulmans* (Pt. 2, Calcutta, 1862), gives a text edition of this treatise, as well as an English translation of its non-modal parts. (The latter are translated in *Temporal Modalities in Arabic Logic*, *op. cit.*).

[18] For Assāl's account of modal propositions see N. Rescher, *Studies in Arabic Philosophy*, *op. cit.*

[19] Shirwānī seems to be following Qazwīnī's text quite closely. Besides there being amazing textual similarities between Shirwānī and Qazwīnī, Shirwānī explicitly refers to *The Sun Epistle* in his discussion of fourth figure moods.

rather than as

(3) It is always true that all men are animals

The difference between (2) and (3) becomes more striking when we consider the modal proposition

(4) $(\sim\forall\mathscr{E})$ (All men are animals)

If we view the qualifying mode here as operating on the categorical proposition "All men are animals" then (4) becomes

(5) It is sometimes true that some men are not animals whereas Shirwānī would have (4) be understood as

(6) All men are not always (i.e., sometimes not) animals. Thus, for example, with regard to the *absolute necessary* $(\Box\mathscr{E})$ proposition and its contradictory, the *general possible* $(\Diamond\mathscr{E})$, the following situation obtains:

| *Original Proposition* | *Contradictory* |
|---|---|
| $(\Box\mathscr{E})$(All $A$ are $B$) = All $A$ are $(\Box\mathscr{E})B$ | Some $A$ are $(\Diamond\mathscr{E})$ not $B$ |
| $(\Box\mathscr{E})$(No $A$ are $B$) = All $A$ are $(\Box\mathscr{E})$ not B | Some $A$ are $(\Diamond\mathscr{E})$ $B$ |
| $(\Box\mathscr{E})$(Some $A$ are $B$) = Some $A$ are $(\Box\mathscr{E})$ B | All $A$ are $(\Diamond\mathscr{E})$ not $B$ |
| $(\Box\mathscr{E})$(Some $A$ are not $B$) = Some $A$ are $(\Box\mathscr{E})$ not $B$ | All $A$ are $(\Diamond\mathscr{E})$ $B$ |

The results of applying the negation principles are set out in Table III.

Table III

CONTRADICTORIES OF SIMPLE MODES

| *Original Proposition* | *Contradictory* |
|---|---|
| $(\Box\mathscr{E})P$ absolute necessary | $(\Diamond\mathscr{E})\sim P$ general possible |
| $(\forall\mathscr{E})P$ absolute perpetual | $(\exists\mathscr{E})\sim P$ general absolute |
| $(\exists\mathscr{E})P$ general absolute | $(\forall\mathscr{E})\sim P$ absolute perpetual |
| $(\Diamond\mathscr{E})P$ general possible | $(\Box\mathscr{E})\sim P$ absolute necessary |
| $(\Box\mathscr{C})P$ general conditional | $(\Diamond\mathscr{C})\sim P$ possible continuing |
| $(\forall\mathscr{C})P$ general conventional | $(\exists\mathscr{C})\sim P$ absolute continuing |
| $(\exists\mathscr{C})P$ absolute continuing | $(\forall\mathscr{C})\sim P$ general conventional |
| $(\Diamond\mathscr{C})P$ possible continuing | $(\Box\mathscr{C})\sim P$ general conditional |
| $(\Box\mathscr{T})P$ absolute temporal | $(\Diamond\mathscr{T})\sim P$ temporal possible |
| $(\mathscr{T})P$ temporal absolute | $(\mathscr{T})\sim P$ temporal absolute |
| $(\Diamond\mathscr{T})P$ temporal possible | $(\Box\mathscr{T})\sim P$ absolute temporal |
| $(\Box\mathscr{S})P$ absolute spread | $(\Diamond\mathscr{S})\sim P$ perpetual (spread) possible |
| $(\mathscr{S})P$ spread absolute | $(\mathscr{S})\sim P$ spread absolute |
| $(\Diamond\mathscr{S})P$ perpetual (spread) possible | $(\Box\mathscr{S})\sim P$ absolute spread |

The situation regarding conversion is more complex. The converse of a modal proposition (X) $P$ is a modal proposition (Y)$P^\circ$ such that

(1) $P^\circ$ is a categorical converse (possibly by limitation) of $P$
(2) (Y)$P^\circ$ is the *strongest modal proposition implied by* (X)$P$.

There seems to be no set procedure for obtaining converses other than the process of demonstration, either by *reductio* or by *supposition*. The results of such conversion demonstrations for the simple modal propositions are listed along with the results for compound modal propositions in Table VI below. We illustrate the conversion procedure with the following examples.

1. $(\Diamond\mathscr{E})$ (All $A$ is $B$) converts to $(\exists\mathscr{C})$ (Some $B$ is $A$). Suppose not. Then, $(\forall\mathscr{C})$ (All $B$ is not $A$), and this, together with the original, yields $(\forall E)$ (All $A$ is not $A$). (For this first figure syllogism see Table VIII below). But this conclusion is a contradiction.
2. $(\exists\mathscr{E})$ (All $A$ is $B$) converts to $(\exists\mathscr{E})$ (Some $B$ is $A$). Let us suppose that $x$ is $A$, then $x$ is $B$ at some time, and hence some $B$ is $A$ at some time. (For a more explicit analysis of this argument, cf. section VI below.)

## 5. The Compound Modes

Compound modes are formed from simple ones by conjoining to the simple modes a *restriction*. This restriction can only take one of the two forms:

1. $(\sim\forall\mathscr{E})$: with non-perpetuity
2. $(\sim\Box\mathscr{E})$: with non-necessity

However, the second form of restriction occurs only as a qualification of basic propositions whose temporality is existential $(\mathscr{E})$.

Taken by themselves, these restrictions qualify the relation of the predicate to the subject in exactly the same way as do the simple modes. For example:

$(\sim\forall\mathscr{E})$ (All $A$ is $B$) $\equiv$ All $A$ is non-perpetually $B$
$(\sim\Box\mathscr{E})$ (All $A$ is $B$) $\equiv$ All $A$ is non-necessarily $B$

Moreover, letting $P'$ be the *contrary* of $P$, we have the general equivalences:

$(\sim\forall\mathscr{E})P \equiv (\exists\mathscr{E})P'$
$(\sim\Box\mathscr{E})P \equiv (\Diamond\mathscr{E})P'$

Note thus the difference between $(\sim\forall\mathscr{E})P$ and $\sim(\forall\mathscr{E})P$, and between $(\sim\Box\mathscr{E})P$ and $\sim(\Box\mathscr{E})P$. (Our previous exegesis errs in suggesting that it is the contradictory rather than the contrary that is at issue.)

The compound modes of categorical propositions are formed by qualifying the relation of the predicate to the subject by a simple mode-*cum*-restriction, so that in a compound mode there is a *two-fold qualification* of the predication relation.

Thus, given a simple mode (X), we are to understand compound modal propositions as follows:

($X \;\&\; \sim\forall\mathscr{E}$)(All $A$ is $B$) ≡ All $A$ is (X) $B$, and they (i.e., those $A$'s that are (X)$B$'s) are not perpetually $B$
≡ All $A$ is (X) $B$, and they are sometimes not $B$

($X \;\&\; \sim\Box\mathscr{E}$)(All $A$ is not $B$) ≡ All $A$ is (X) not $B$, and they are not necessarily not $B$
≡ All $A$ is (X) not $B$, and they are possibly $B$

The situation regarding the other categorical forms that are not displayed here is entirely analogous. Thus, for example, the general absolute ($\exists\mathscr{E}$) can be compounded into the non-perpetual existential ($\exists\mathscr{E} \;\&\; \sim\forall\mathscr{E}$) or the non-necessary existential ($\exists\mathscr{E} \;\&\; \sim\Box\mathscr{E}$):

($\exists\mathscr{E} \;\&\; \sim\forall\mathscr{E}$)(Some $A$ is $B$) ≡ Some $A$ is sometimes $B$, and they are not always $B$
≡ Some $A$ is sometimes $B$, and they sometimes not $B$

($\exists\mathscr{E} \;\&\; \sim\Box\mathscr{E}$)(Some $A$ is $B$) ≡ Some $A$ is sometimes $B$, and they are not necessarily $B$
≡ Some $A$ is sometimes $B$, and they are possibly not $B$

Some further examples of compound modal propositions are:

($\Box\mathscr{C} \;\&\; \sim\forall\mathscr{E}$): All writers move of necessity as long as they write, but not perpetually

($\Box\mathscr{T} \;\&\; \sim\forall\mathscr{E}$): All moons are of necessity not eclipsed at the time of the quarter moon, but not perpetually

($\Diamond\mathscr{E} \;\&\; \sim\Box\mathscr{E}$): With a special possibility, all fires are cold

Note here that an affirmative (negative) compound modal proposition is composed of an affirmative (negative) simple modal proposition and a negative (affirmative) general absolute or general possible.

The compound modes presented by Al-Shirwānī are set forth in Table IV. It does not appear that he considered this list to be exhaustive of all the obtainable compound modes. His considerations seem to have centered around those compound modes which were needed for conversion and—above all—for first figure syllogisms. (See Table

VIII below.) For example, those compounds of possibility which are conspicuously absent in Table IV are presumably missing because they *invariably* yield non-productive syllogisms in the first figure.[20]

Table IV

COMPOUND MODES IN SHIRWANI[21]

| *Type* | *Name* |
|---|---|
| I. MODES OF NECESSITY | |
| $\Box\mathscr{E}$ & $\sim\mathscr{E}$ | non-perpetual necessary (*) (impossible combination) |
| $\Box\mathscr{C}$ & $\sim\forall\mathscr{E}$ | special conditional |
| $\Box\mathscr{T}$ & $\sim\forall\mathscr{E}$ | temporal |
| $\Box\mathscr{S}$ & $\sim\forall\mathscr{E}$ | spread |
| II. MODES OF PERPETUALITY | |
| $\forall\mathscr{E}$ & $\sim\forall\mathscr{E}$ | non-perpetual perpetual (*) (impossible combination) |
| $\forall\mathscr{C}$ & $\sim\forall\mathscr{E}$ | special conventional |
| III. MODES OF ACTUALITY | |
| $\exists\mathscr{E}$ & $\sim\forall\mathscr{E}$ | non-perpetual existential |
| $\exists\mathscr{E}$ & $\sim\Box\mathscr{E}$ | non-necessary existential |
| $\exists\mathscr{C}$ & $\sim\forall\mathscr{E}$ | non-perpetual continuing (*) |
| $\mathscr{T}$ & $\sim\forall\mathscr{E}$ | non-perpetual temporal absolute (*) |
| $\mathscr{S}$ & $\sim\forall\mathscr{E}$ | non-perpetual spread absolute (*) |
| IV. MODES OF POSSIBILITY | |
| $\Diamond\mathscr{E}$ & $\sim\Box\mathscr{E}$ | special possible[22] |
| $\Diamond\mathscr{E}$ & $\sim\forall\mathscr{E}$ | non-necessary existential[23] |

(*): Missing in Qazwīnī.

[20] In fact, when Shirwānī introduces the compound modes he specifically says that there are (only) *eight* of them, which he goes on to discuss, namely, ($\Box\mathscr{C}$ & $\sim\forall\mathscr{E}$), ($\forall\mathscr{C}$ & $\sim\forall\mathscr{E}$), ($\Box\mathscr{T}$ & $\sim\forall\mathscr{E}$), ($\Box\mathscr{S}$ & $\sim\forall\mathscr{E}$), ($\exists\mathscr{C}$ & $\sim\forall\mathscr{E}$), ($\exists\mathscr{E}$ & $\sim\forall\mathscr{E}$), ($\exists\mathscr{E}$ & $\sim\Box\mathscr{E}$), and ($\Diamond\mathscr{E}$ & $\sim\Box\mathscr{E}$). Later in the text, however, he mentions four additional modes. The matter seems to resolve itself if we consider the eight modes in question to be the *standard* modes "into which it is usual to inquire."

[21] Concerning this classification of modes, it should be mentioned that neither Qazwīnī nor Shirwānī present a classification of modes as such. Rather, they refer to modes as follows. Propositions are either *actuals* or *possibles*. The actuals consist of the perpetuals, $\Box\mathscr{E}$, $\forall\mathscr{E}$; the conditionals, $\Box\mathscr{C}$, $\Box\mathscr{C}$ & $\sim\forall\mathscr{E}$; the conventionals, $\forall\mathscr{C}$, $\forall\mathscr{C}$ & $\sim\forall\mathscr{E}$; the continuing propositions, $\exists\mathscr{C}$, $\exists\mathscr{C}$ & $\sim\forall\mathscr{E}$; the conventionals, $\forall\mathscr{C}$, $\forall\mathscr{C}$ & $\sim\forall\mathscr{E}$; the continuing propositions, $\exists\mathscr{C}$, $\exists\mathscr{C}$ & $\sim\forall\mathscr{E}$; the temporals, $\Box\mathscr{T}$, $\mathscr{T}$, $\Box\mathscr{S}$, $\mathscr{S}$, $\Box\mathscr{T}$ & $\sim\forall\mathscr{E}$, $\Box\mathscr{S}$ & $\sim\forall\mathscr{E}$; the existentials, $\exists\mathscr{E}$ & $\sim\forall\mathscr{E}$, $\exists\mathscr{E}$ & $\sim\Box\mathscr{E}$; and the general absolute, $\exists\mathscr{E}$. The possibles are $\Diamond\mathscr{C}$, $\Diamond\mathscr{T}$, $\Diamond\mathscr{S}$, $\Diamond\mathscr{E}$, and $\Diamond\mathscr{E}$ & $\sim\Box\mathscr{E}$.

[22] Note here that $(\Diamond\mathscr{E} \,\&\, \sim\Box\mathscr{E})P \equiv (\Diamond\mathscr{E} \,\&\, \sim\Box\mathscr{E})P'$, where $P'$ is the contrary of P. Thus it is described in the text as "composed of two general possibles, one negative, the other positive."

[23] Note that $\Diamond\mathscr{E}$ & $\sim\forall\mathscr{E}$ is the equivalent with $\exists\mathscr{E}$ & $\sim\Box\mathscr{E}$ and is thus the non-necessary existential all over again.

## 6. The Negation and Conversion of the Compound Modes

The negation of a compound mode follows the negation of each of its component modes in the following way:

> Given an affirmative (negative) compound modal proposition, which is an affirmative (negative) simple modal proposition *conjoined* with a negative (affirmative) general absolute, or a negative (affirmative) general possible—its negation is the negation of the simple modal proposition *disjoined* with an affirmative (negative) absolute perpetual, or an affirmative (negative) absolute perpetual, or an affirmative (negative) absolute necessary.

For example,

$\sim(\forall\mathscr{C}\ \&\ \forall\mathscr{E})$(All $A$ is $B$) $\equiv$ Some $A$ is not $B$ while they are $A$, or they are perpetually $B$

$\sim(\Box\mathscr{S}\ \&\ \sim\forall\mathscr{E})$(All $A$ is $B$) $\equiv$ Some $A$ is not $B$ possibly at all times, or they are perpetually $B$

Let us introduce some notation which will enable us to adequately describe the negation process for compound modes. Given a categorical proposition $P$ let us define $P^*$, the pronomialization of $P$, as follows:

| $P$ | $P^*$ |
|---|---|
| All $A$ is $B$ | *they* (i.e., the $A$'s at issue) are $B$ |
| All $A$ is not $B$ | *they* (i.e., the $A$'s at issue) are not $B$ |
| Some $A$ is $B$ | *they* (i.e., the $A$'s at issue) are $B$ |
| Some $A$ is not $B$ | *they* (i.e., the $A$'s at issue) are not $B$ |

Table V

Contradictories of Compound Modes

| *Compound Original* | *Contradictory* |
|---|---|
| $(\Box\mathscr{E}\ \&\ \sim\forall\mathscr{E})P$ | $(\Diamond\mathscr{E})\sim P \vee (\forall\mathscr{E})P^*$ |
| $(\Box\mathscr{C}\ \&\ \sim\forall\mathscr{E})P$ | $(\Diamond\mathscr{C})\sim P \vee (\forall\mathscr{E})P^*$ |
| $(\Box\mathscr{T}\ \&\ \sim\forall\mathscr{E})P$ | $(\Diamond\mathscr{T})\sim P \vee (\forall\mathscr{E})P^*$ |
| $(\Box\mathscr{S}\ \&\ \sim\forall\mathscr{E})P$ | $(\Diamond\mathscr{S})\sim P \vee (\forall\mathscr{E})P^*$ |
| $(\forall\mathscr{E}\ \&\ \sim\forall\mathscr{E})P$ | $(\exists\mathscr{E})\sim P \vee (\forall\mathscr{E})P^*$ |
| $(\forall\mathscr{C}\ \&\ \sim\forall\mathscr{E})P$ | $(\exists\mathscr{C})\sim P \vee (\forall\mathscr{E})P^*$ |
| $(\exists\mathscr{C}\ \&\ \sim\forall\mathscr{E})P$ | $(\forall\mathscr{C})\sim P \vee (\forall\mathscr{E})P^*$ |
| $(\mathscr{T}\ \&\ \sim\forall\mathscr{E})P$ | $(\mathscr{T})\sim P \vee (\forall\mathscr{E})P^*$ |
| $(\mathscr{S}\ \&\ \sim\forall\mathscr{E})P$ | $(\forall\mathscr{E})\sim P \vee (\forall\mathscr{E})P^*$ |
| $(\exists\mathscr{E}\ \&\ \sim\forall\mathscr{E})P$ | $(\forall\mathscr{E})\sim P \vee (\forall\mathscr{E})P^*$ |
| $(\exists\mathscr{E}\ \&\ \sim\Box\mathscr{E})P$ | $(\forall\mathscr{E})\sim P \vee (\Box\mathscr{E})P^*$ |
| $(\Diamond\mathscr{E}\ \&\ \sim\forall\mathscr{E})P$ | $(\Box\mathscr{E})\sim P \vee (\forall\mathscr{E})P^*$ |
| $(\Diamond\mathscr{E}\ \&\ \sim\Box\mathscr{E})P$ | $(\Box\mathscr{E})\sim P \vee (\Box\mathscr{E})P^*$ |

We can now represent a compound mode $(\text{X} \mathbin{\&} {\sim}\forall\mathscr{E})$, or $(\text{X} \mathbin{\&} {\sim}\Box\mathscr{E})$ as follows:

$$(\text{X} \mathbin{\&} {\sim}\forall\mathscr{E})P \equiv ((\text{X})P \mathbin{\&} ({\sim}\forall\mathscr{E})P^*$$

$$(\text{X} \mathbin{\&} {\sim}\Box\mathscr{E})P \equiv (\text{X})P \mathbin{\&} ({\sim}\Box\mathscr{E})P^*$$

and negation can be described as in Table V.

The conversion process for compound modal propositions is essentially analogous to that for the simple modes. Given a compound mode $(\text{X} \mathbin{\&} {\sim}\text{Y})P$, its converse, if there is one, is a proposition $(\text{Z})P^\circ$, where (Z) can be either simple or compound, such that

1. $P^\circ$ is the categorical converse of $P$
2. $(\text{Z})P^\circ$ is the strongest modal proposition such that $(\text{X} \mathbin{\&} {\sim}\text{Y})P$ implies $(\text{Z})P^\circ$

Again, as for simple modes, the procedure for determining conversion is that of *demonstration*. Thus, for example, the special conventional converts to the non-perpetual absolute continuing, in the universal affirmative case:

> $(\forall\mathscr{C} \mathbin{\&} {\sim}\forall\mathscr{E})$ (All *A* is *B*) converts to $(\exists\mathscr{C} \mathbin{\&} {\sim}\forall\mathscr{E})$ (Some *B* is *A*). Otherwise, $(\forall\mathscr{C})$ (All *B* is not *A*) or $(\forall\mathscr{E})$ (they are *A*); so that $(\forall\mathscr{C})$ (All *B* is not *A*) or $(\forall\mathscr{E})$ (All *B* is *A*). But the first disjunct together with the original yields $(\forall\mathscr{C})$ (All *A* is not *A*), which is absurd, and the second disjunct together with the original simple mode yields $(\forall\mathscr{E})$ (All *B* is *B*), which, with the original restriction $(\exists\mathscr{E})$ (All *B* is not *B*), in turn yields a pair of contradictories. (For these first figure syllogisms see Table VIII below.)

In Qazwīnī and Shirwānī there are three references to the "non-perpetual-about-some conventional": by Qazwīnī in *The Sun Epistle* §67/65 and §72/70, and by Shirwānī in the present text in Table XIA for the sixth mood (AEE) of the fourth figure. In §67/65 Qazwīnī says that the universal negative general conditional and general conventional convert to the universal negative general conventional, and that the universal negative special conditional and special conventional convert to the *non-perpetual-about-some conventional.*

> The reason of this process in reference to the general conventional is that it is an *adherent* of both kinds of general propositions. The reason why the converted proposition is non-perpetual-about-some is because [if] it is not true that some *B* is with general absoluteness *C*, [then] it is true by perpetuity that no *B* is *C*, and thus [this is] converted into perpetually no *C* is *B*. But the original proposition was [that no *C* is *B* as long as it is *C* but not perpetually, and so] that every *C* is *B*.

Now, the (partial) converse of $(\forall\mathscr{C}\ \&\ \sim\forall\mathscr{E})$ (No $C$ is $B$) that *adheres* to the generals is $(\forall\mathscr{C})$ (No $B$ is $C$). Also, $(\forall\mathscr{E})$ (No $C$ is $B$), which is All $C$ is sometimes $B$, converts to Some $B$ is sometimes $C$, which is $(\sim\forall\mathscr{E})$ (Some $B$ is not $C$). Thus the converse of $(\forall\mathscr{C}\ \&\ \sim\forall\mathscr{E})$ (All $C$ is not $B$) is $(\forall\mathscr{C})$ (All $B$ is not $C$) & $(\sim\forall\mathscr{E})$ (Some $B$ is not $C$); and the latter is aptly called "the non-perpetual-about-some conventional." We note, thus, that (1) *only a universal negative* special proposition converts to a non-perpetual-about-some conventional; and (2) the latter is sort of a half-way house between the universal negative specials and the particular negative specials—yet it is neither of them.

---

Table VI

CONVERSION OF MODAL PROPOSITIONS ACCORDING TO CATEGORICAL FORM

| | | A, I | E | O |
|---|---|---|---|---|
| | *Original* | *Converse* | *Converse* | *Converse* |
| | $(\square\mathscr{E})P$ | $(\exists\mathscr{C})P^\circ$ | $(\forall\mathscr{E})P^\circ_d$ | — |
| | $(\forall\mathscr{E})P$ | $(\exists\mathscr{C})P^\circ$ | $(\forall\mathscr{E})P^\circ_d$ | — |
| | $(\square\mathscr{C})P$ | $(\exists\mathscr{C})P^\circ$ | $(\forall\mathscr{C})P^\circ_d$ | — |
| | $(\forall\mathscr{C})P$ | $(\exists\mathscr{C})P^\circ$ | $(\forall\mathscr{C})P^\circ_d$ | — |
| (c) | $(\exists\mathscr{C})P$ | $(\exists\mathscr{C})P^\circ$ | — | — |
| (c) | $(\square\mathscr{T})P$ | $(\text{E}\mathscr{E})P^\circ$ | — | — |
| (c) | $(\square\mathscr{S})P$ | $(\exists\mathscr{E})P^\circ$ | — | — |
| (c) | $(\mathscr{T})P$ | $(\exists\mathscr{E})P^\circ$ | — | — |
| (c) | $(\mathscr{S})P$ | $(\exists\mathscr{E})P^\circ$ | — | — |
| | $(\exists\mathscr{E})P$ | $(\exists\mathscr{E})P^\circ$ | — | — |
| | $(\Diamond\mathscr{X})P$ | (— *) | — | — |
| (c) | $(\square\mathscr{E}\ \&\sim\forall\mathscr{E})P$ | $(\exists\mathscr{C}\ \&\sim\forall\mathscr{E})P^\circ$ | $(\forall\mathscr{E})P^\circ_d\ \&(\sim\forall\mathscr{E})P^\circ_i$ (@) | — |
| (c) | $(\forall\mathscr{E}\ \&\sim\forall\mathscr{E})P$ | $(\exists\mathscr{C}\ \&\sim\forall\mathscr{E})P^\circ$ | $(\forall\mathscr{E})P^\circ_d\ \&(\sim\forall\mathscr{E})P^\circ_i$ | — |
| | $(\square\mathscr{C}\ \&\sim\forall\mathscr{E})P$ | $(\exists\mathscr{C}\ \&\sim\forall\mathscr{E})P^\circ$ | $(\forall\mathscr{C})P^\circ_d\ \&(\sim\mathscr{E})P^\circ_i$ (#) | $(\forall\mathscr{C}\ \&\sim\forall\mathscr{E})P^\circ$ |
| | $(\forall\mathscr{C}\ \&\sim\forall\mathscr{E})P$ | $(\exists\mathscr{C}\ \&\sim\forall\mathscr{E})P^\circ$ | $(\forall\mathscr{C})P^\circ_d\ \&(\sim\forall\mathscr{E})P^\circ_i$ | $(\forall\mathscr{C}\ \&\sim\forall\mathscr{E})P^\circ$ |
| (c) | $(\exists\mathscr{C}\ \&\sim\forall\mathscr{E})P$ | $(\exists\mathscr{C})P^\circ$ | — | — |
| | $(\square\mathscr{T}\ \&\sim\forall\mathscr{E})P$ | $(\exists\mathscr{E})P^\circ$ | — | — |
| | $(\square\mathscr{S}\ \&\sim\forall\mathscr{E})P$ | $(\exists\mathscr{E})P^\circ$ | — | — |
| (c) | $(\mathscr{T}\ \&\sim\forall\mathscr{E})P$ | $(\exists\mathscr{E})P^\circ$ | — | — |
| (c) | $(\mathscr{S}\ \&\sim\forall\mathscr{E})P$ | $(\exists\mathscr{E})P^\circ$ | — | — |
| | $(\exists\mathscr{E}\ \&\sim\forall\mathscr{E})P$ | $(\exists\mathscr{E})P^\circ$ | — | — |
| | $(\exists\mathscr{E}\ \&\sim\square\mathscr{E})P$ | $(\exists\mathscr{E})P^\circ$ | — | — |
| | $(\Diamond\mathscr{E}\ \&\sim\mathscr{X})P$ | — | — | — |

(c) Our own calculations.
(#): The non-perpetual-about-some conventional.
(*) Not convertible in this categorical form.
(@): The non-perpetual-about-some perpetual, constructed in analogy with the non-perpetual-about-some conventional.

---

This interpretation is further verified by the fact that Shirwānī has the non-perpetual-about-some conventional as a conclusion for the mood **AEE–4**, where the syllogism does not yield a particular special proposition.

The conversion results for both simple and compound modes is given in Table VI above. These results are not given by Shirwānī, but are taken from Qazwīnī, and Qazwīnī's account is supplemented by our own calculations. (In Table VI we represent the *direct* converse of a universal negative proposition $P$, i.e., the converse *not* by limitation, as $P^{\circ}_{\mathrm{d}}$, and the converse of $P$ by limitation as $P^{\circ}_{l}$.)

## 7. The Logical Analysis of Modal Propositions

We shall now attempt an analysis of the modal propositions considered thus far in terms of presentday symbolic notation. $\mathrm{R}_t$ is the basic operator for realization-at-time-$t$.[24] We shall make use of the following abbreviations:

| | | |
|---|---|---|
| $\mathscr{T}\mathrm{Q}x = \mathrm{R}_{\mathscr{T}}(\mathrm{Q}x)$ | $\Box\mathscr{T}\mathrm{Q}x = \Box\mathrm{R}_{\mathscr{T}}(\mathrm{Q}x)$ | $\Diamond\mathscr{T}\mathrm{Q}x = \Diamond\mathrm{R}_{\mathscr{T}}(\mathrm{Q}x)$ |
| $\mathscr{S}\mathrm{Q}x = \mathrm{R}_{\mathscr{S}}(\mathrm{Q}x)$ | $\Box\mathscr{S}\mathrm{Q}x = \Box\mathrm{R}_{\mathscr{S}}(\mathrm{Q}x)$ | $\Diamond\mathscr{S}\mathrm{Q}x = \Diamond\mathrm{R}_{\mathscr{S}}(\mathrm{Q}x)$ |
| $\exists\mathrm{Q}x = (\exists t)\mathrm{R}_t(\mathrm{Q}x)$ | $\exists\Box\mathrm{Q}x = (\exists t)\Box\mathrm{R}_t(\mathrm{Q}x)$ | $\exists\Diamond\mathrm{Q}x = (\exists \bar{t})\Diamond\mathrm{R}_t(\mathrm{Q}x)$ |
| $\forall\mathrm{Q}x = (\forall t)\mathrm{R}_t\mathrm{Q}x)$ | $\forall\Box\mathrm{Q}x = (\forall t)\Box\mathrm{R}_t(\mathrm{Q}x)$ | $\forall\Diamond\mathrm{Q}x = (\forall t)\Diamond\mathrm{R}_t(\mathrm{Q}x)$ |

In our symbolizations of modal propositions, *we shall systematically suppress the temporality condition* ($\mathscr{E}$) relating to the existence of the subject.

Concerning the symbolic rendition of modes, we take notice of just the following points. First, in adopting the symbolic machinery that we have, we assume here that all the usual quantificational and modal principles hold. Secondly, in the $\mathscr{E}$-modes the existence condition has been suppressed; fully stated, ($\Box\mathscr{E}$) (All $A$ is $B$), for example, would be $(\forall x)\ [(\exists\, t)\ \mathrm{R}_t\ Ax \supset (\forall t)\ \Box\mathrm{R}_t\ ((Ex \supset Bx)]$. Thirdly, the modes $\mathscr{T}$ and $\mathscr{S}$ are special time-instantiations, with regard to the existence of the subject, and accordingly, we here use '$\mathscr{T}$' and '$\mathscr{S}$' as time-constants.

Since the texts have very little to say about the implicational relations among modes, we must rely heavily on our symbolic interpretation of the modes to be able to say what relations hold. There is, in particular, a question concerning the relationship between the $\mathscr{T}$-modes and the $\mathscr{C}$-modes. As an example of the temporal absolute ($\mathscr{T}$) Shirwānī gives "All writers move at the time they are writing." Comparing this with the continuing absolute ($\exists\mathscr{C}$) "All writers move

[24] For details regarding this operator see N. Rescher and A. Urquhart, *Temporal Logic* (New York, and Vienna, 1971).

*Simple Modes of the* **A** *Proposition (All A is B)*

| *Type* | *Example* | *Name* |
|---|---|---|
| $(\Box\mathscr{E})$ | $(\forall x)[\exists Ax \supset \forall\Box Bx]$ | absolute necessary |
| $(\Box\mathscr{C})$ | $(\forall x)[\exists Ax \supset \forall\Box(Ax \supset Bx)]$ | general conditional |
| $(\Box\mathscr{T})$ | $(\forall x)[\exists Ax \supset \Box\mathscr{T}Bx]$ | absolute temporal |
| $(\Box\mathscr{S})$ | $(\forall x)[\exists Ax \supset \Box\mathscr{S}Bx]$ | absolute spread |
| $(\forall\mathscr{E})$ | $(\forall x)[\exists Ax \supset \forall Bx]$ | absolute perpetual |
| $(\forall\mathscr{C})$ | $(\forall x)[\exists Ax \supset \forall(Ax \supset Bx)]$ | general conventional |
| $(\exists\mathscr{C})$ | $(\forall x)[\exists Ax \supset \exists(Ax \,\&\, Bx)]$ | absolute continuing |
| $(\mathscr{T})$ | $(\forall x)[\exists Ax \supset \mathscr{T}Bx]$ | temporal absolute |
| $(\mathscr{S})$ | $(\forall x)[\exists Ax \supset \mathscr{S}Bx]$ | spread absolute |
| $(\exists\mathscr{E})$ | $(\forall x)[\exists Ax \supset \exists Bx]$ | general absolute |
| $(\Diamond\mathscr{C})$ | $(\forall x)[\exists Ax \supset \exists\Diamond(Ax \,\&\, Bx)]$ | possible continuing |
| $(\Diamond\mathscr{T})$ | $(\forall x)[\exists Ax \supset \Diamond\mathscr{T}Bx]$ | temporal possible |
| $(\Diamond\mathscr{E})$ | $(\forall x)[\exists Ax \supset \exists\Diamond Bx]$ | general possible |
| $(\Diamond\mathscr{S})$ | $(\forall x)[\exists Ax \supset \Diamond\mathscr{S}Bx]$ | perpetual possible |

*Compound Modes of the* **A** *Proposition (All A is B)*

|  |  |  |
|---|---|---|
| $(\Box\mathscr{E} \,\&\, \sim\forall\mathscr{E})$ | $(\forall x)\{\exists Ax \supset [\forall\Box Bx \,\&\, \sim\forall Bx]\}$ | non-perpetual necessary |
| $(\Box\mathscr{C} \,\&\, \sim\forall\mathscr{E})$ | $(\forall x)\{\exists Ax \supset [\forall\Box(Ax \supset Bx) \,\&\, \sim\forall Bx]\}$ | special conditional |
| $(\Box\mathscr{T} \,\&\, \sim\forall\mathscr{E})$ | $(\forall x)\{\exists Ax \supset [\Box\mathscr{T}Bx \,\&\, \sim\forall Bx]\}$ | temporal |
| $(\Box\mathscr{S} \,\&\, \sim\forall\mathscr{E})$ | $(\forall x)\{\exists Ax \supset [\Box\mathscr{S}Bx \,\&\, \sim\forall Bx]\}$ | spread |
| $(\forall\mathscr{E} \,\&\, \sim\forall\mathscr{E})$ | $(\forall x)\{\exists Ax \supset [\forall Bx \,\&\, \sim\forall Bx]\}$ | non-perpetual perpetual |
| $(\forall\mathscr{C} \,\&\, \sim\forall\mathscr{E})$ | $(\forall x)\{\exists Ax \supset [\forall Ax \supset Bx) \,\&\, \sim\forall Bx]\}$ | special conventional |
| $(\exists\mathscr{C} \,\&\, \sim\forall\mathscr{E})$ | $(\forall x)\{\exists Ax \supset [\exists(Ax \,\&\, Bx) \,\&\, \sim\forall Bx]\}$ | non-perpetual continuing absolute |
| $(\mathscr{T} \,\&\, \sim\forall\mathscr{E})$ | $(\forall x)\{\exists Ax \supset [\mathscr{T}Bx \,\&\, \sim\forall Bx]\}$ | non-perpetual temporal. absolute |
| $(\mathscr{S} \,\&\, \sim\forall\mathscr{E})$ | $(\forall x)\{\exists Ax \supset [\mathscr{S}Bx \,\&\, \sim\forall Bx]\}$ | non-perpetual spread absolute |
| $(\exists\mathscr{E} \,\&\, \sim\forall\mathscr{E})$ | $(\forall x)\{\exists Ax \supset [Bx \,\&\, \sim\forall Bx]\}$ | non-perpetual existential |
| $(\exists\mathscr{E} \,\&\, \sim\Box\mathscr{E})$ | $(\forall x)\{\exists Ax \supset [\exists Bx \,\&\, \sim\forall\Box Bx]\}$ | non-necessary existential |
| $(\Diamond\mathscr{E} \,\&\, \sim\Box\mathscr{E})$ | $(\forall x)\{\exists Ax \supset [\exists\Diamond Bx \,\&\, \sim\forall\Box Bx]\}$ | special possible |

while they are writing," we would conclude that Shirwānī holds that $\exists\mathscr{C}\rightarrow\mathscr{T}$. Table VII, then, presents the implicational relations among modes as calculated from the given constructions. Note that the compound modes (X & $\sim\forall\mathscr{E}$) and (X & $\sim\Box\mathscr{E}$) both imply the simple mode (X).

Table VII

RELATIVE STRENGTHS OF MODAL PROPOSITIONS

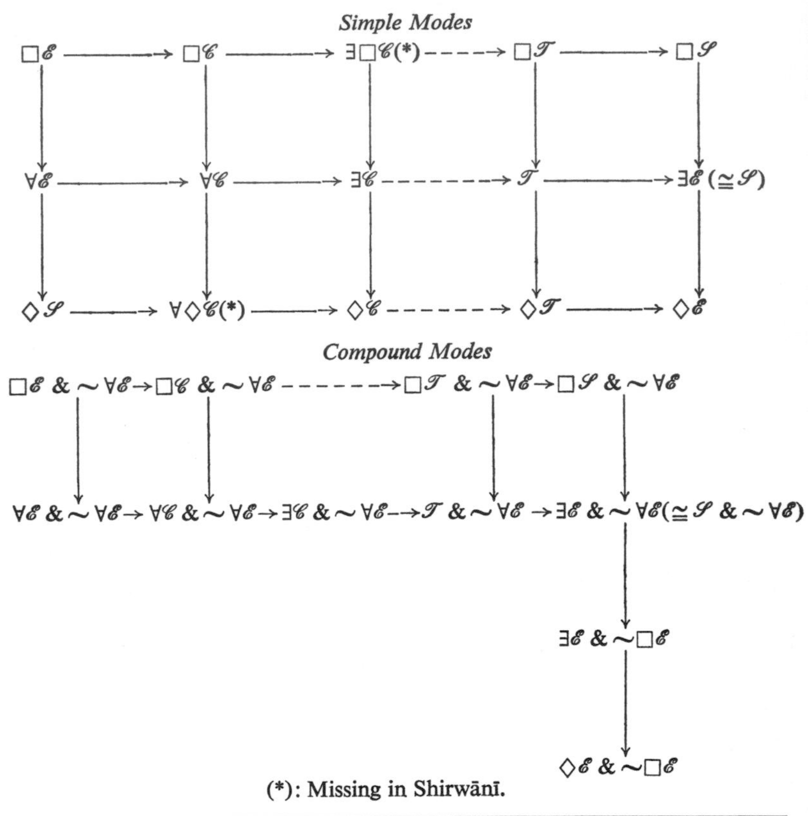


(*): Missing in Shirwānī.

When we view modes in the light of the symbolic apparatus just presented, it becomes clear that we may distinguish five additional modes. In Section 8 we will see that these five modes are necessary to describe adequately the third figure syllogisms. It was noted that the mode ($\mathscr{T}$) is really a time-instantiation with respect to the temporality ($\mathscr{E}$). On analogy, we can also have a time-instantiation with respect to the temporality ($\mathscr{C}$), thus giving rise to three new modes:

($\Box\mathscr{T}\mathscr{C}$) as $(\forall x)\,[(\exists t)\,R_t Ax \supset \Box \mathrm{R}_{\mathscr{T}}(Ax \,\&\, Bx)]$ continuing absolute temporal

($\mathscr{T}\mathscr{C}$) as $(\forall x)\,[(\exists t)\,\mathrm{R}_t Ax \supset \mathrm{R}_{\mathscr{T}}\,(Ax \,\&\, Bx)]$ continuing temporal absolute

($\Diamond\mathscr{T}\mathscr{C}$) as $(\forall x)\,[(\exists t)\,R_t Ax \supset \Diamond \mathrm{R}_{\mathscr{T}}(Ax \,\&\, Bx)$ continuing temporal possible

Also, since the temporality ($\mathscr{S}$) is tantamount to the modality ($\exists$) combined with temporality ($\mathscr{E}$), we can, on analogy with the modes ($\Box\mathscr{S}$) and ($\Diamond\mathscr{S}$), distinguish the modes:

($\Box\mathscr{S}\mathscr{C}$) as $(\forall x)\,[(\exists t)\,R_t Ax \supset (\exists t)\Box R_t\,(Ax\ \&\ Bx)]$
continuing absolute spread

($\Diamond\mathscr{S}\mathscr{C}$) as $(\forall x)\,[(\exists t)\,R_t Ax \supset (\forall t)\Diamond R_t\,(Ax\ \&\ Bx)]$
continuing perpetual possible

The compound modes that could be constructed out of the new modes are to be construed on analogy with the other compounds, for example:

($\Box\mathscr{T}\mathscr{C}\ \&\sim\forall\mathscr{E}$) as $(\forall x)\{(\exists t)R_t Ax \supset (\Box R_{\mathscr{T}}(Ax\ \&\ Bx)\ \&\sim(\forall t)R_t Bx]\}$

To make the relation of the new modes to the other modes clearer, we present in Table VIIA the relative strengths of the augmented number of simple modes. We shall in this table explicitly display the modalities ($\forall$) and ($\exists$) and the temporality ($\mathscr{E}$) that are implicitly present in certain modes.

Table VIIA

RELATIVE STRENGTHS OF MODAL PROPOSITIONS

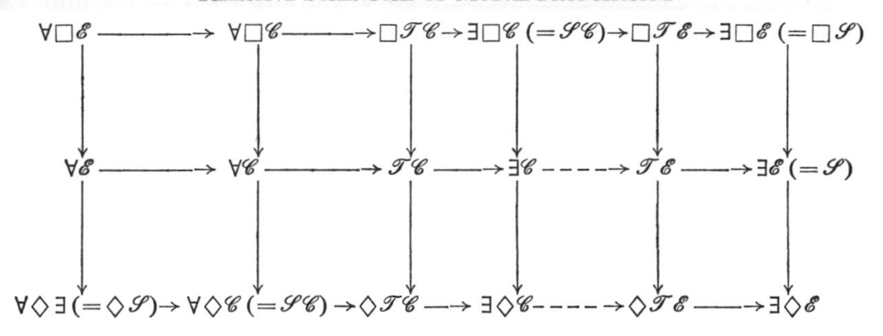


## 8. First Figure Syllogisms

In "The Sun Epistle" *Al-Risālah al-shamsiyyah* (§81), al-Qazwīnī specifies the productive (i.e., valid) first figure modal syllogisms as follows:

> As to the first figure, its condition regarding modality is the actuality of the minor.[25] The conclusion here is the same as the major, if

[25] In the earlier translation of §81, in *Temporal Modalities in Arabic Logic*, the first sentence was erroneously translated as "as to the first figure, its condition [obtains] in relation to the modality operative for the minor."

> it (i.e., the major)[26] is other than one of the two conditionals and the two conventionals; and otherwise (i.e., if the major *is* one of these four) it is like the minor when it is without the condition of the non-necessary or the non-perpetual, and the necessity which belongs specially (i.e., only) to the minor,[27] if the major is one of the two generals, and adding the non-perpetual to it if it is one of the two specials.

Shirwānī renders the last clause more clearly, and says:

> otherwise, it is like the minor, omitting the non-necessary, the non-perpetual, and the necessity special to it, if it was found in it, adding the non-perpetual of the major, if it was found in it.

Thus, the account for the first figure syllogisms is the following:

(1) The minor premiss must be one of the seventeen actuals.[28]
(2) If the major is not one of ($\Box\mathscr{C}$), ($\forall\mathscr{C}$), ($\Box\mathscr{C}$ & $\sim\forall\mathscr{E}$), and ($\forall\mathscr{C}$ & $\sim\forall\mathscr{E}$), then the mode of the conclusion is that of the major.
(3) If the major is one of these four, then the mode of the conclusion is like that of the minor except that
  (a) the *restriction* of the conclusion is the same as the restriction of the major
  (b) the conclusion is necessitated if and only if both the minor and the major are.
(4) All other moods are non-productive.

Shirwānī supplements this account by a table. Table VIII gives Shirwānī's table as he himself presents it (using, of course, the mode names, rather than our symbolic abbreviations). Note that the table deals only with condition (3). Concerning the first figure Shirwānī says that there are four valid categorical moods (*Barbara*, *Celarent*, *Darii*, and *Ferio*) and that each of these four, when mixed with modes, give rise to 374 productive moods,[29] resulting from the seventeen

[26] In the earlier translation of §81, *op. cit.*, the clause "if it is other than one of the two conditionals and the two conventionals" was interpreted as "if it (i.e., the minor) is other than. . . ." In the light of Shirwānī's account, however, the present interpretation is clearly the correct one.

[27] In the earlier translation of §81, *op. cit.*, the phrase "and the necessity which belongs specially to the minor" was (erroneously) suppressed as a seeming corruption of the text.

[28] Note that there are twenty-two possible major, and minor, premisses: fourteen simples and eight compounds, which divide into seventeen actuals and five possibles. Thus, for example, there are seventeen minor premisses—the actuals—displayed in Table VIII, since the possibles as minor are non- productive.

[29] Note, thus, that when these 374 modal moods are combined with the four categorical moods there results a total of 1946 productive syllogistic moods in the first figure alone!

Table VIII (♯)

FIRST FIGURE SYLLOGISMS

| | major / minor | $\Box\mathscr{C}$ | $\forall\mathscr{C}$ | $\Box\mathscr{C} \,\&\! \sim \forall\mathscr{E}$ | $\forall\mathscr{C} \,\&\! \sim \forall\mathscr{E}$ |
|---|---|---|---|---|---|
| 1 | $\Box\mathscr{E}$ | $\Box\mathscr{E}$ | $\forall\mathscr{E}$ | $\Box\mathscr{E} \,\&\! \sim \forall\mathscr{E}$ | $\forall\mathscr{E} \,\&\! \sim \forall\mathscr{E}$ |
| 2 | $\forall\mathscr{E}$ | $\forall\mathscr{E}$ | $\forall\mathscr{E}$ | $\forall\mathscr{E} \,\&\! \sim \forall\mathscr{E}$ | $\forall\mathscr{E} \,\&\! \sim \forall\mathscr{E}$ |
| 3 | $\Box\mathscr{C}$ | $\Box\mathscr{C}$ | $\forall\mathscr{C}$ | $\Box\mathscr{C} \,\&\! \sim \forall\mathscr{C}$ | $\forall\mathscr{C} \,\&\! \sim \forall\mathscr{E}$ |
| 4 | $\Box\mathscr{T}$ | $\Box\mathscr{T}$ | $\mathscr{T}$ | $\Box\mathscr{T} \,\&\! \sim \forall\mathscr{E}$ | $\mathscr{T} \,\&\! \sim \forall\mathscr{E}$(*) |
| 5 | $\Box\mathscr{S}$ | $\Box\mathscr{S}$ | $\mathscr{S}$ | $\Box\mathscr{S} \,\&\! \sim \forall\mathscr{E}$ | $\mathscr{S} \,\&\! \sim \forall\mathscr{E}$(*) |
| 6 | $\forall\mathscr{C}$ | $\forall\mathscr{C}$ | $\forall\mathscr{C}$ | $\forall\mathscr{C} \,\&\! \sim \forall\mathscr{E}$ | $\forall\mathscr{C} \,\&\! \sim \forall\mathscr{E}$ |
| 7 | $\exists\mathscr{C}$ | $\exists\mathscr{C}$ | $\exists\mathscr{C}$ | $\exists\mathscr{C} \,\&\! \sim \forall\mathscr{E}$ | $\exists\mathscr{C} \,\&\! \sim \forall\mathscr{E}$ |
| 8 | $\mathscr{T}$ | $\mathscr{T}$ | $\mathscr{T}$ | $\mathscr{T} \,\&\! \sim \forall\mathscr{E}$ | $\mathscr{T} \,\&\! \sim \forall\mathscr{E}$ |
| 9 | $\mathscr{S}$ | $\mathscr{S}$ | $\mathscr{S}$ | $\mathscr{S} \,\&\! \sim \forall\mathscr{E}$ | $\mathscr{S} \,\&\! \sim \forall\mathscr{E}$ |
| 10 | $\Box\mathscr{C} \,\&\! \sim \forall\mathscr{E}$ | $\Box\mathscr{C}$ | $\forall\mathscr{C}$ | $\Box\mathscr{C} \,\&\! \sim \forall\mathscr{E}$ | $\forall\mathscr{C} \,\&\! \sim \forall\mathscr{E}$(*) |
| 11 | $\forall\mathscr{C} \,\&\! \sim \forall\mathscr{E}$ | $\forall\mathscr{C}$ | $\forall\mathscr{C}$ | $\forall\mathscr{C} \,\&\! \sim \forall\mathscr{E}$ | $\forall\mathscr{C} \,\&\! \sim \forall\mathscr{E}$ |
| 12 | $\Box\mathscr{T} \,\&\! \sim \forall\mathscr{E}$ | $\Box\mathscr{T}$ | $\mathscr{T}$ | $\Box\mathscr{T} \,\&\! \sim \forall\mathscr{E}$ | $\mathscr{T} \,\&\! \sim \forall\exists$(*) |
| 13 | $\Box\mathscr{S} \,\&\! \sim \forall\mathscr{E}$ | $\Box\mathscr{S}$ | $\mathscr{S}$ | $\Box\mathscr{S} \,\&\! \sim \forall\mathscr{E}$ | $\mathscr{S} \,\&\! \sim \forall\mathscr{E}$(*) |
| 14 | $\exists\mathscr{E} \,\&\! \sim \forall\mathscr{E}$ | $\exists\mathscr{E}$ | $\exists\mathscr{E}$ | $\exists\mathscr{E} \,\&\! \sim \forall\mathscr{E}$ | $\exists\mathscr{E} \,\&\! \sim \forall\mathscr{E}$ |
| 15 | $\exists\mathscr{C} \,\&\! \sim \forall\mathscr{E}$ | $\exists\mathscr{C}$ | $\exists\mathscr{C}$ | $\exists\mathscr{C} \,\&\! \sim \forall\mathscr{E}$ | $\exists\mathscr{C} \,\&\! \sim \forall\mathscr{E}$ |
| 16 | $\exists\mathscr{E} \,\&\! \sim \Box\mathscr{E}$ | $\exists\mathscr{E}$ | $\exists\mathscr{E}$ | $\exists\mathscr{E} \,\&\! \sim \forall\mathscr{E}$ | $\exists\mathscr{E} \,\&\! \sim \forall\mathscr{E}$ |
| 17 | $\exists\mathscr{E}$ | $\exists\mathscr{E}$ | $\exists\mathscr{E}$ | $\exists\mathscr{E} \,\&\! \sim \forall\mathscr{E}$ | $\exists\mathscr{E} \,\&\! \sim \forall\mathscr{E}$ |

(♯): Displayed by Shirwānī.

(*): Proposed correction, in accordance with condition (3b), removing the necessity of the simple component mode.

actual minor modes times the twenty-two major modes—the fourteen simples and the eight standard compounds $(\Box\mathscr{C}\ \&\ {\sim}\forall\mathscr{E})$, $(\forall\mathscr{C}\ \&\ {\sim}\forall\mathscr{E})$, $(\exists\mathscr{C}\ \&\ {\sim}\forall\mathscr{E})$, $(\Box\mathscr{T}\ \&\ {\sim}\forall\mathscr{E})$, $(\Box\mathscr{S}\ \&\ {\sim}\forall\mathscr{E})$, $(\exists\mathscr{E}\ \&\ {\sim}\forall\mathscr{E})$, $\exists\mathscr{E}\ \&\ {\sim}\Box\mathscr{E})$, and $(\Diamond\mathscr{E}\ \&\ {\sim}\Box\mathscr{E})$.[30]

As far as we can determine, the standard account given by both Qazwīnī and Shirwānī is correct except for moods containing the *continuing* modes $(\exists\mathscr{C})$, $(\exists\mathscr{C}\ \&\ {\sim}\forall\mathscr{E})$, and $(\Diamond\mathscr{C})$ in the major. In these first figure moods, as far as can be ascertained by independent calculation, the conclusion is $(\exists\mathscr{E})$, $(\exists\mathscr{E}\ \&\ {\sim}\forall\mathscr{E})$, and $(\Diamond\mathscr{E})$, respectively, for each of the seventeen minors.

The account of modal syllogisms in the first figure duly corrected in the indicated manner can be verified by means of the symbolic apparatus for modal propositions given below in Section 6. It is intended to show by the following examples of Fitch-style deductions that the various claims regarding syllogistic results are in fact justified.

*Example* 1

major: $(\Box\mathscr{E})$(All $B$ are $C$)
minor: $(\exists\mathscr{E}\ \&\ {\sim}\forall\mathscr{E})$(All $A$ are $B$)
conclusion: $(\Box\mathscr{E})$(All $A$ are $C$)

| | | |
|---|---|---|
| 1 | $(\forall x)[\exists Bx \supset \forall\Box Cx]$ | |
| 2 | $(\forall x)[\exists Ax \supset (\exists Bx\ \&\ {\sim}\forall Bx)]$ | |
| 3 | $\exists Ax$ | |
| 4 | $\exists Bx\ \&\ {\sim}\forall Bx$ | 2, 3 |
| 5 | $\exists Bx$ | 4 |
| 6 | $\forall\Box Cx$ | 1, 5 |
| 7 | $(\forall x)[\exists Ax \supset \forall\Box Cx]$ | 3–6 |

*Example* 2

major: $(\Box\mathscr{C}\ \&\ {\sim}\forall\mathscr{E})$(All $B$ are $C$)
minor: $(\exists\mathscr{E})$(All $A$ are $B$)
conclusion: $(\exists\mathscr{E}\ \&\ {\sim}\forall\mathscr{E})$(All $A$ are $C$)

| | | |
|---|---|---|
| 1 | $(\forall x)[\exists Bx \supset (\forall\Box(Bx \supset Cx)\ \&\ {\sim}\forall Cx)]$ | |
| 2 | $(\forall x)[\exists Ax \supset \exists Bx]$ | |
| 3 | $\exists Ax$ | |
| 4 | $\exists Bx$ | 2, 3 |
| 5 | $\forall\Box(Bx \supset Cx)\ \&\ {\sim}\forall Cx$ | 1, 4 |
| 6 | $\forall\Box(Bx \supset Cx)$ | 5 |
| 7 | $\exists Cx$ | 4, 6 |
| 8 | $\exists Cx\ \&\ {\sim}\forall Cx$ | 5, 7 |
| 9 | $(\forall x)[\exists Ax \supset (\exists Cx\ \&\ {\sim}\forall Cx)]$ | 3–8 |

*Example* 3

major: $(\exists\mathscr{C})$(All $B$ are $C$)
minor: $(\forall\mathscr{E})$(All $A$ are $B$)
conclusion: $(\exists\mathscr{E}\ \&\ {\sim}\forall\mathscr{E})$(All $A$ are $C$)

| | | |
|---|---|---|
| 1 | $(\forall x)[\exists Bx \supset \exists(Bx\ \&\ Cx)]$ | |
| 2 | $(\forall x)[\exists Ax \supset \forall(Ax \supset Bx)]$ | |
| 3 | $\exists Ax$ | |
| 4 | $\forall(Ax \supset Bx)$ | 2, 3 |
| 5 | $\exists Bx$ | 3, 4 |
| 6 | $\exists(Bx\ \&\ Cx)$ | 1, 5 |
| 7 | $\exists Cx$ | 6 |
| 8 | $(\forall x)[\exists Ax \supset \exists Cx]$ | 3–7 |

[30] See footnote 12, section IV. Note also that the other four compounds never occur as premisses in Shirwānī's account of the four figure, and that the eight standard compounds but for $(\exists\mathscr{C}\ \&\ {\sim}\forall\mathscr{E})$ are the only seven compounds discussed by Qazwīnī.

Thus, for modal syllogisms in Shirwānī we have neither the Aristotelian rule of inference regarding modal syllogistics that the mode of the conclusion follows the mode of the major, since, as in Example 2, it sometimes follows the mode of the minor; nor the variant rule that it follows the minor, since, as in Example 1, it sometimes follows the major; nor the Theophrastean (*peiorem*) rule that it follows the mode of the weaker premiss, since, as in Example 1 it sometimes follows the mode of the stronger. Moreover, as in Example 3 the mode of the conclusion sometimes follows neither the mode of the major nor the minor. Note also that the restriction of the conclusion mode follows only the restriction of the major mode, as is illustrated in Examples 1 and 2. Note finally, as in Example 1, that when the major does not involve the temporality ($\mathscr{C}$), the Aristotelian rule that the conclusion follows the major *does* obtain. In general, however, the logical situation in the theory of temporalized modal syllogistic of the Arab logicians is far more subtle than in the Aristotelian tradition of their Greek precursors.

### 9. Second, Third, and Fourth Figure Syllogisms

The first figure syllogisms were held to be self-evident, and the other syllogisms were to be demonstrated by reduction to the first figure by converting one or both premisses, by interchanging the premisses and converting the conclusion, or by *reductio ad impossibile*.

Concerning the second figure Shirwānī says that each of the four categorical moods (*Cesare*, *Camestres*, *Festino*, and *Baroco*), when combined with modes, gives rise to 144 productive moods.[31] Shirwānī's account, in perfect accord with Qazwīnī,[32] is as follows.

There are two conditions for valid syllogisms in the second figure: (1) truth by perpetuity must pertain to the minor (so that the minor is $\Box\mathscr{E}$ or $\forall\mathscr{E}$), or the major must be one of the convertible negative propositions (i.e., one of $\Box\mathscr{E}$, $\forall\mathscr{E}$, $\Box\mathscr{C}$, $\forall\mathscr{C}$, $\Box\mathscr{C}$ & $\sim\forall\mathscr{E}$, and $\forall\mathscr{C}$ & $\sim\forall\mathscr{E}$); (2) a possibility proposition may be used only when the other premiss is necessary (and is, thus, $\Box\mathscr{E}$, $\Box\mathscr{C}$, or $\Box\mathscr{C}$ & $\sim\forall\mathscr{E}$). If both these conditions are met, then: if either premiss is perpetually true, then the conclusion is $\forall\mathscr{E}$; if the major is a conditional or a conventional proposition, then the mode of the conclusion

[31] The 2 perpetual minors times 17 actuals majors, plus the remaining 15 actual minors times the 6 negative convertible majors, plus the absolute necessary minor times the 5 possible majors, plus the absolute necessary major times the 5 possible minors, plus the 2 conditional majors times the 5 possible minors—144 moods.

[32] And so it is with all four figures: Shirwānī is in accord with Qazwīnī throughout.

Table IX(*)

SECOND FIGURE SYLLOGISMS

<table>
<tr><th colspan="2">major<br>minor</th><th>$\Box\mathscr{C}$</th><th>$\Box\mathscr{C}$ & $\sim\forall\mathscr{E}$</th><th>$\forall\mathscr{C}$</th><th>$\forall\mathscr{C}$ & $\sim\forall\mathscr{E}$</th></tr>
<tr><td>1</td><td>$\Box\mathscr{C}$</td><td colspan="4" rowspan="4">$\forall\mathscr{C}$</td></tr>
<tr><td>2</td><td>$\forall\mathscr{C}$</td></tr>
<tr><td>3</td><td>$\Box\mathscr{C}$ & $\sim\forall\mathscr{E}$</td></tr>
<tr><td>4</td><td>$\forall\mathscr{C}$ & $\sim\forall\mathscr{E}$</td></tr>
<tr><td>5</td><td>$\exists\mathscr{E}$</td><td colspan="4">$\exists\mathscr{E}$</td></tr>
<tr><td>6</td><td>$\exists\mathscr{C}$</td><td colspan="4">$\exists\mathscr{C}$</td></tr>
<tr><td>7</td><td>$\mathscr{T}$</td><td colspan="4">$\mathscr{T}$</td></tr>
<tr><td>8</td><td>$\mathscr{S}$</td><td colspan="4">$\mathscr{S}$</td></tr>
<tr><td>9</td><td>$\exists\mathscr{E}$ & $\sim\forall\mathscr{E}$</td><td colspan="4" rowspan="2">$\exists\mathscr{E}$</td></tr>
<tr><td>10</td><td>$\exists\mathscr{E}$ & $\sim\Box\mathscr{E}$</td></tr>
<tr><td>11</td><td>$\exists\mathscr{C}$ & $\sim\forall\mathscr{E}$</td><td colspan="4">$\exists\mathscr{C}$</td></tr>
<tr><td>12</td><td>$\Box\mathscr{T}$ & $\sim\forall\mathscr{E}$</td><td colspan="4" rowspan="2">$\mathscr{T}$</td></tr>
<tr><td>13</td><td>$\Box\mathscr{T}$</td></tr>
<tr><td>14</td><td>$\Box\mathscr{S}$ & $\sim\forall\mathscr{E}$</td><td colspan="4" rowspan="2">$\mathscr{S}$</td></tr>
<tr><td>15</td><td>$\Box\mathscr{S}$</td></tr>
<tr><td>16</td><td>$\Diamond\mathscr{E}$</td><td colspan="2" rowspan="2">$\Diamond\mathscr{E}$</td><td colspan="2" rowspan="5">non-productive</td></tr>
<tr><td>17</td><td>$\Diamond\mathscr{E}$ & $\sim\Box\mathscr{E}$</td></tr>
<tr><td>18</td><td>$\Diamond\mathscr{C}$</td><td colspan="2">$\Diamond\mathscr{C}$</td></tr>
<tr><td>19</td><td>$\Diamond\mathscr{T}$</td><td colspan="2">$\Diamond\mathscr{T}$</td></tr>
<tr><td>20</td><td>$\Diamond\mathscr{S}$</td><td colspan="2">$\Diamond\mathscr{S}$</td></tr>
</table>

(*): Displayed by Shirwānī.

is like that of the minor except without restriction and without necessity. All remaining moods are non-productive.

In Table IX we reproduce Shirwānī's table for the second figure. This table concerns only the case when the major is a conditional or a conventional proposition. The remaining cases are as just described.

As regards the third figure, there are six valid categorical moods (*Darapti*, *Felapton*, *Datisi*, *Ferison*, *Disamis*, and *Bocardo*), each producing 374 valid modal moods. The account given by Shirwānī is the following.

The condition for syllogisms in the third figure is that the minor premiss be one of the actual propositions. When the major is a conditional or a conventional proposition, the mode of the conclusion is like the mode of the converse of the minor, removing the non-perpetual from it or adding it to it, according as the major is general or special. Otherwise, the mode of the conclusion is like that of the major. All other moods are non-productive.

Table X is the table that Shirwānī presents for the third figure. And this table is correct in its entirety. The undisplayed cases, however, present some difficulty, in that the given account seems not to adequately describe them.[33] As far as can be ascertained by independent calculation, the given account is correct except (1) when the major is a *continuing* mode, the conclusion is like the major with ($\mathscr{C}$) *weakened* to ($\mathscr{E}$) in certain places, and (2) when the minor is true by perpetuity, the conclusion is like the major with ($\mathscr{E}$) *strengthened* to ($\mathscr{C}$) in certain places. Specifically, the situation is as follows.

When the major is $\Box\mathscr{C}$, $\forall\mathscr{C}$, $\Box\mathscr{C}$ & $\sim\forall\mathscr{E}$, $\forall\mathscr{C}$ & $\sim\forall\mathscr{E}$, the mode of the conclusion is like the mode of the converted minor removing the non-perpetual from it or adding it to it, according as the major is general or special. When the major is $\exists\mathscr{C}$, $\exists\mathscr{C}$ & $\sim\forall\mathscr{E}$ the conclusion is like the major, with ($\mathscr{C}$) weakened to ($\mathscr{E}$) in all cases in which the modality ($\forall$) does not pertain to the minor. When the major is $\Diamond\mathscr{C}$, the conclusion is like the major with ($\mathscr{C}$) weakened to ($\mathscr{E}$) in all cases in which the modality ($\forall\Box$) does not pertain to the minor.

Otherwise, when the major is not one of these seven, the following holds. When the minor is ($\Box\mathscr{E}$), the conclusion is like the major, with ($\mathscr{E}$) strengthened to ($\mathscr{C}$) in all cases for which ($\forall$) does not pertain to the major. When the minor is ($\forall\mathscr{E}$), the conclusion is like the major, with ($\mathscr{E}$) strengthened to ($\mathscr{C}$) in all cases for which ($\forall$), ($\Box$), or

[33] For example, according to the text we are to have $(\exists\mathscr{E})P$, $(\forall\mathscr{E})P'$, therefore, $(\exists\mathscr{E})P''$. Yet, it is clear that the following holds: $(\forall x)[\exists Mx \supset \exists Px]$, $(\exists x)[\exists Mx$ & $\forall Sx]$, therefore, $(\exists x)[\exists Sx$ & $\exists(Sx$ & $Px)]$. If our interpretation of modes is correct, the mood should thus be $(\exists\mathscr{E})P$, $(\forall\mathscr{E})P'$, therefore, $(\exists\mathscr{C})P''$.

Table X(*)

THIRD FIGURE SYLLOGISMS

<table>
<tr><th colspan="2">minor \ major</th><th>$\Box\mathscr{C}$</th><th>$\forall\mathscr{C}$</th><th>$\Box\mathscr{C}$ & $\sim\forall\mathscr{E}$</th><th>$\forall\mathscr{C}$ & $\sim\forall\mathscr{E}$</th></tr>
<tr><td>1</td><td>$\Box\mathscr{E}$</td><td colspan="2" rowspan="8">$\exists\mathscr{C}$</td><td colspan="2" rowspan="8">$\exists\mathscr{C}$ & $\forall\mathscr{E}$</td></tr>
<tr><td>2</td><td>$\forall\mathscr{E}$</td></tr>
<tr><td>3</td><td>$\Box\mathscr{C}$</td></tr>
<tr><td>4</td><td>$\Box\mathscr{C}$ & $\sim\forall\mathscr{E}$</td></tr>
<tr><td>5</td><td>$\forall\mathscr{C}$</td></tr>
<tr><td>6</td><td>$\forall\mathscr{C}$ & $\sim\forall\mathscr{E}$</td></tr>
<tr><td>7</td><td>$\exists\mathscr{C}$ & $\sim\forall\mathscr{E}$</td></tr>
<tr><td>8</td><td>$\exists\mathscr{E}$</td></tr>
<tr><td>9</td><td>$\Box\mathscr{T}$ & $\sim\forall\mathscr{E}$</td><td colspan="2">$\exists\mathscr{E}$(♯)</td><td colspan="2">$\exists\mathscr{E}$ & $\sim\forall\mathscr{E}$(♯)</td></tr>
<tr><td>10</td><td>$\Box\mathscr{S}$ & $\sim\forall\mathscr{E}$</td><td colspan="2" rowspan="8">$\exists\mathscr{E}$</td><td colspan="2" rowspan="8">$\exists\mathscr{E}$ & $\sim\forall\mathscr{E}$</td></tr>
<tr><td>11</td><td>$\Box\mathscr{T}$</td></tr>
<tr><td>12</td><td>$\Box\mathscr{S}$</td></tr>
<tr><td>13</td><td>$\mathscr{T}$</td></tr>
<tr><td>14</td><td>$\mathscr{S}$</td></tr>
<tr><td>15</td><td>$\exists\mathscr{E}$ & $\sim\forall\mathscr{E}$</td></tr>
<tr><td>16</td><td>$\exists\mathscr{E}$ & $\sim\forall\mathscr{E}$</td></tr>
<tr><td>17</td><td>$\exists\mathscr{E}$</td></tr>
</table>

(*) Displayed by Shirwānī.
(♯): Here Shirwānī has ($\exists\mathscr{C}$) and ($\exists\mathscr{C}$ & $\sim\forall\mathscr{E}$).

($\Diamond$) do not pertain to the major. Otherwise (when the minor is neither ($\Box\mathscr{E}$) nor ($\forall\mathscr{E}$)), the conclusion is like the major. (Note that we here have need of the new modes introduced in Section 6.)

Finally, as to the fourth figure, Shirwānī says that there are eight valid moods, five of which are categorically valid (*Bramantip*, *Dimaris*, *Camenes*, *Fesapo*, and *Fresison*), the other three (**AOO**, **OAO**, and **IEO**) being valid only when the negative premiss is one of the specials. Shirwānī, noting his departure from Qazwīnī, orders the moods as follows: (i) **AII**, (ii) **IAI**, (iii) **EAO**, (iv) **OAO**, (v) **EIO**, (vi) **AEE**, (vii) **AOO**, and (viii) **IEO**. The first, second, sixth, and eighth mood are reduced by interchanging the premisses and converting the (resultant) conclusion; the third and the fifth by converting each of the premisses; the seventh by converting the minor, resulting in a third figure syllogism.

The conditions for fourth figure syllogisms as given by Shirwānī are as follows. (1) Both premisses must be actuals. (2) The negative propositions in the syllogism must be convertible. (3) In the sixth mood (**AEE**) the minor must be true by perpetuity (or else the major mode must be one of the six negative convertibles).[34] (4) In the seventh

Table XI(*)

CERTAIN FOURTH FIGURE SYLLOGISMS

<table>
<tr><th>minor<br>major</th><th>$\Box\mathcal{E}$</th><th>$\forall\mathcal{E}$</th><th>$\Box\mathcal{C}$</th><th>$\forall\mathcal{C}$</th><th>$\Box\mathcal{C}$ & $\sim\forall\mathcal{E}$</th><th>$\forall\mathcal{C}$ & $\sim\forall\mathcal{E}$</th><th>remaining actuals</th></tr>
<tr><td>$\Box\mathcal{E}$</td><td colspan="4" rowspan="6">$\exists\mathcal{C}$</td><td colspan="2" rowspan="2">(♯)</td><td rowspan="8">$\exists\mathcal{E}$</td></tr>
<tr><td>$\forall\mathcal{E}$</td></tr>
<tr><td>$\Box\mathcal{C}$</td><td colspan="2" rowspan="4">$\exists\mathcal{C}$ & $\sim\forall\mathcal{E}$</td></tr>
<tr><td>$\forall\mathcal{C}$</td></tr>
<tr><td>$\Box\mathcal{C}$ & $\sim\forall\mathcal{E}$</td></tr>
<tr><td>$\forall\mathcal{C}$ & $\sim\forall\mathcal{E}$</td></tr>
<tr><td>$\exists\mathcal{C}$</td><td colspan="2" rowspan="3"></td><td colspan="4" rowspan="2">$\exists\mathcal{C}$(**)</td></tr>
<tr><td>$\exists\mathcal{C}$ & $\sim\forall\mathcal{E}$</td></tr>
<tr><td>remaining actuals</td><td colspan="5">$\exists\mathcal{E}$</td></tr>
</table>

(*): Not displayed, but only described by Shirwānī.
(♯): Since the premisses here are contradictory, the conclusion is problematic.
(**): Described by Shirwānī as ($\exists\mathcal{E}$).

[34] This clause is missing in the text, but Qazwīnī's otherwise identical discussion contains the clause. Cf. Table XIA.

mood (**AOO**), the major mode must be one of the six negative convertibles. (5) In the eighth mood (**IEO**) the minor must be one of the two specials, and the major one of the negative convertibles.

The productive combinations in both the first and second moods (**AAI, IAI**) are 289. Their condition is that the premisses be actual propositions. The mode of the conclusion is the converse of the minor, if the minor is $\Box\mathscr{E}$ or $\forall\mathscr{E}$, or if both premisses are in the six negative convertibles. Otherwise, the conclusion mode is $(\exists\mathscr{E})$. All other cases are non-productive. This situation is presented in Table XI.

The productive combinations in both the third and fifth moods (**EAO, EIO**) are 102. Their condition is the general condition that the premisses be actual and that the negative premiss be convertible. The moods are reduced to the first figure by converting each premiss. Thus, the conclusion is $\forall\mathscr{E}$, if the major is $\Diamond\mathscr{E}$ or $\forall\mathscr{E}$; otherwise, the mode of the conclusion is the same as the mode of the converted minor after removing the non-perpetual from it. All other cases are non-productive. The situation is presented in Table XIA.

The productive combinations in the fourth mood (**OAO**) are 34. The condition for the fourth mood is the general condition that the premisses be actual and that the negative premiss be convertible and, therefore, that the major be one of the specials. The conclusion is the same as in the third figure after converting the major. Since the major is a special proposition, the conclusion is thus the same as the converse of the minor. All other cases are non-productive. The situation is as displayed in Table XIB.

The productive combinations in the sixth mood (**AEE**) are 58. The condition for the sixth mood is the general condition that the negative premiss (the minor) is one of the negative convertibles, and the particular condition that the minor is $\Box\mathscr{E}$ or $\forall\mathscr{E}$, or that the major is a negative convertible. The conclusion mode is $\forall\mathscr{E}$, if either premiss is $\Box\mathscr{E}$ or $\forall\mathscr{E}$; otherwise, the conclusion has the same mode as the converse of the minor. All other cases are non-productive. The situation is as displayed in Table XIC.

The productive combinations in both the seventh and eighth moods (**AOO, IEO**) are 12. The condition for the seventh mood (**AOO**) is the particular condition that the major is one of the negative convertibles, and the general condition that the negative premiss be convertible and, thus, that the minor premiss is one of the specials. The mood is reduced to the second figure by converting the minor. All other cases are non-productive. The situation is displayed in Table XID.

Table XIA(*)

CERTAIN FOURTH FIGURE SYLLOGISMS

| | major / minor | $\Box\mathscr{E}$ | $\forall\mathscr{E}$ | $\Box\mathscr{C}$ | $\forall\mathscr{C}$ | $\Box\mathscr{C}$ & $\sim\forall\mathscr{E}$ | $\forall\mathscr{C}$ & $\sim\forall\mathscr{E}$ |
|---|---|---|---|---|---|---|---|
| 1 | $\Box\mathscr{E}$ | | | | | | |
| 2 | $\forall\mathscr{E}$ | | | | | | |
| 3 | $\Box\mathscr{C}$ | | | | | | |
| 4 | $\forall\mathscr{C}$ | | | | | | |
| 5 | $\forall\mathscr{C}$ & $\sim\forall\mathscr{E}$ | | | | | | |
| 6 | $\Box\mathscr{C}$ & $\sim\forall\mathscr{E}$ | | | | | | |
| 7 | $\exists\mathscr{C}$ & $\sim\forall\mathscr{E}$ | | | | | | |
| 8 | $\exists\mathscr{C}$ | | | | | $\exists\mathscr{C}$(#) | |
| 9 | $\Box\mathscr{T}$ & $\sim\forall\mathscr{E}$ | $\forall\mathscr{E}$ | | | | | |
| 10 | $\Box\mathscr{S}$ & $\sim\forall\mathscr{E}$ | | | | | | |
| 11 | $\Box\mathscr{T}$ | | | | | | |
| 12 | $\Box\mathscr{S}$ | | | | $\exists\mathscr{E}$ | | |
| 13 | $\mathscr{T}$ | | | | | | |
| 14 | $\mathscr{S}$ | | | | | | |
| 15 | $\exists\mathscr{E}$ & $\sim\Box\mathscr{E}$ | | | | | | |
| 16 | $\exists\mathscr{E}$ & $\sim\forall\mathscr{E}$ | | | | | | |
| 17 | $\exists\mathscr{E}$ | | | | | | |

(*): Displayed by Shirwānī.

(#): Shirwānī has ($\exists\mathscr{E}$) here.

Table XIB(*)

CERTAIN FOURTH FIGURE SYLLOGISMS

<table>
<tr><th colspan="2">minor \ major</th><th>$\Box\mathscr{C}$ & $\sim\forall\mathscr{E}$</th><th>$\forall\mathscr{C}$ & $\sim\forall\mathscr{E}$</th></tr>
<tr><td>1</td><td>$\Box\mathscr{E}$</td><td colspan="2" rowspan="8">$\exists\mathscr{C}$ & $\sim\forall\mathscr{E}$</td></tr>
<tr><td>2</td><td>$\forall\mathscr{E}$</td></tr>
<tr><td>3</td><td>$\Box\mathscr{C}$</td></tr>
<tr><td>4</td><td>$\forall\mathscr{C}$</td></tr>
<tr><td>5</td><td>$\Box\mathscr{C}$ & $\sim\forall\mathscr{E}$</td></tr>
<tr><td>6</td><td>$\forall\mathscr{C}$ & $\sim\forall\mathscr{E}$</td></tr>
<tr><td>7</td><td>$\exists\mathscr{C}$ & $\sim\forall\mathscr{E}$</td></tr>
<tr><td>8</td><td>$\exists\mathscr{C}$</td></tr>
<tr><td>9</td><td>$\Box\mathscr{T}$</td><td colspan="2">$\exists\mathscr{E}$ & $\sim\forall\mathscr{E}$ (#)</td></tr>
<tr><td>10</td><td>$\Box\mathscr{S}$</td><td colspan="2" rowspan="8">$\exists\mathscr{E}$ & $\sim\forall\mathscr{E}$</td></tr>
<tr><td>11</td><td>$\Box\mathscr{T}$ & $\sim\forall\mathscr{E}$</td></tr>
<tr><td>12</td><td>$\Box\mathscr{S}$ & $\sim\forall\mathscr{E}$</td></tr>
<tr><td>13</td><td>$\mathscr{T}$</td></tr>
<tr><td>14</td><td>$\mathscr{S}$</td></tr>
<tr><td>15</td><td>$\exists\mathscr{E}$ & $\sim\forall\mathscr{E}$</td></tr>
<tr><td>16</td><td>$\exists\mathscr{E}$ & $\sim\forall\mathscr{E}$</td></tr>
<tr><td>17</td><td>$\exists\mathscr{E}$</td></tr>
</table>

(*) Displayed by Shirwānī.

(#): Shirwānī has $\forall\mathscr{C}$ & $\sim\forall\mathscr{E}$ here.

Table XIC (*)

CERTAIN FOURTH FIGURE SYLLOSIGMS

<table>
<tr><th colspan="2">minor<br>major</th><th>$\Box\mathscr{E}$</th><th>$\forall\mathscr{E}$</th><th>$\Box\mathscr{C}$</th><th>$\forall\mathscr{C}$</th><th>$\Box\mathscr{C}$ & $\forall\mathscr{E}$</th><th>$\forall\mathscr{C}$ & $\sim\forall\mathscr{E}$</th></tr>
<tr><td>1</td><td>$\Box\mathscr{E}$</td><td rowspan="17" colspan="2">$\forall\mathscr{E}$</td><td rowspan="2" colspan="2">$\forall\mathscr{E}$ (♯)</td><td rowspan="2" colspan="2">$\forall\mathscr{E}$</td></tr>
<tr><td>2</td><td>$\forall\mathscr{E}$</td></tr>
<tr><td>3</td><td>$\Box\mathscr{C}$</td><td rowspan="4" colspan="2">$\forall\mathscr{C}$</td><td rowspan="4" colspan="2">the "non-perpetual-about-some conventional"</td></tr>
<tr><td>4</td><td>$\forall\mathscr{C}$</td></tr>
<tr><td>5</td><td>$\Box\mathscr{C}$ & $\sim\forall\mathscr{E}$</td></tr>
<tr><td>6</td><td>$\forall\mathscr{C}$ & $\sim\forall\mathscr{E}$</td></tr>
<tr><td>7</td><td>$\Box\mathscr{S}$ & $\sim\forall\mathscr{E}$</td><td rowspan="11" colspan="4">non-productive</td></tr>
<tr><td>8</td><td>$\Box\mathscr{T}$ & $\sim\forall\mathscr{E}$</td></tr>
<tr><td>9</td><td>$\Box\mathscr{T}$</td></tr>
<tr><td>10</td><td>$\Box\mathscr{S}$</td></tr>
<tr><td>11</td><td>$\mathscr{T}$</td></tr>
<tr><td>12</td><td>$\mathscr{S}$</td></tr>
<tr><td>13</td><td>$\exists\mathscr{C}$ & $\sim\forall\mathscr{E}$</td></tr>
<tr><td>14</td><td>$\exists\mathscr{C}$</td></tr>
<tr><td>15</td><td>$\exists\mathscr{E}$ & $\sim\forall\mathscr{E}$</td></tr>
<tr><td>16</td><td>$\exists\mathscr{E}$ & $\sim\Box\mathscr{E}$</td></tr>
<tr><td>17</td><td>$\exists\mathscr{E}$</td></tr>
</table>

(*): Displayed by Shirwānī.

(#): Shirwānī has ($\forall\mathscr{C}$) here.

Table XID(*)

CERTAIN FOURTH FIGURE SYLLOGISMS

| major \ minor | | $\Box\mathscr{C}$ & $\sim\forall\mathscr{E}$ | $\forall\mathscr{C}$ & $\sim\forall\mathscr{E}$ |
|---|---|---|---|
| 1 | $\Box\mathscr{E}$ | $\forall\mathscr{E}$ | |
| 2 | $\forall\mathscr{E}$ | | |
| 3 | $\Box\mathscr{C}$ | $\forall\mathscr{C}$ | |
| 4 | $\forall\mathscr{C}$ | | |
| 5 | $\Box\mathscr{C}$ & $\sim\forall\mathscr{E}$ | | |
| 6 | $\forall\mathscr{C}$ & $\sim\forall\mathscr{E}$ | | |

(*): Displayed by Shirwānī.

Table XIE(*)

CERTAIN FOURTH FIGURE SYLLOGISMS

| major \ minor | | $\Box\mathscr{C}$ & $\sim\forall\mathscr{E}$ | $\forall\mathscr{C}$ & $\sim\forall\mathscr{E}$ |
|---|---|---|---|
| 1 | $\Box\mathscr{E}$ | $\forall\mathscr{C}$ & $\sim\forall\mathscr{E}$(♯) | |
| 2 | $\forall\mathscr{E}$ | | |
| 3 | $\Box\mathscr{C}$ | $\forall\mathscr{C}$ & $\sim\forall\mathscr{E}$ | |
| 4 | $\forall\mathscr{C}$ | | |
| 5 | $\Box\mathscr{C}$ & $\sim\forall\mathscr{E}$ | | |
| 6 | $\forall\mathscr{C}$ & $\sim\forall\mathscr{E}$ | | |

(*): Displayed by Shirwānī.
(♯): Since the premisses are contradictory here, the conclusion is problematic.

The condition for the eighth mood (**IEO**) is the particular condition that the major be one of the negative convertibles and that the minor be one of the specials. The mood is reduced to the first figure by interchanging the premisses and converting the conclusion. All other cases are non-productive. The situation is displayed in Table XIE.

With the end of the fourth figure, ends Shirwānī's account of temporal modal syllogisms, which, despite its occasional slips, reveals a detailed and sophisticated comprehension of temporal modal logic.

## 10. Temporal Modalities Among the Ancient Greeks

The early history of temporal quantifiers like "sometimes" and "always," and of the theory of temporalized modalities is that linked to them through the mediation of such principles as "What is sometimes actual is always possible" remains shrouded in obscurity. We know that the rudiments of such a theory was actively developed by the ancient Greeks: the Megarians and the Stoics,[35] and Aristotle and the early Peripatetics.[36] The notions of temporalized modality that are at work here are mainly those relating to the "Master Argument" of Diodorus Cronus.[37]

There seems to have been a disagreement as to modality between the Stoics and the Megarians. On the Megarian view:

(1) The *actually true* is that which is actually realized *now*, so that, using $R_t$ as operator for chronological realization, so that "$R_t(P)$" stands for "$P$ is realized at the time $t$,"

$$T_n(P) \text{ iff } R_n(P)$$

with n = *now*, or more generally

$$T_t(P) \text{ iff } R_t(P)$$

(2) The *possible* is that which is actually realized (i.e., true) at some present-or-future time

$$P_n(P) \text{ iff } (\exists t)\ [t \geq n\ \&\ R_n(P)]$$

or more generally

$$P_t(P) \text{ iff } (\exists t')[t' \geq t\ \&\ R_{t'}(P)]$$

[35] See E. Zeller, *Die Philosophie der Griechen*, pt. 3, vol. I (5th ed., Leipzig, 1923); and B. Mates, *Stoic Logic* (Berkeley and Los Angeles, 1953), see esp. pp. 36–41.

[36] I. M. Bochenski, *La logique de Théophraste* (Freiburg, 1947).

[37] See Mates, *op. cit.*, pp. 38–39. Cf. J. Hintikka, "Aristotle and the 'Master Argument' of Diodorus," *American Philosophical Quarterly*, vol. I (1964), pp. 101–114. And see also N. Rescher, "A Version of the 'Master Argument' of Diodorus," *The Journal of Philosophy*, vol. 63 (1966), pp. 438–445.

(3) The necessary is that which is actually realized at every future time

$$N_n(p) \text{ iff } (\forall t)[t \geq n \rightarrow R_n(P)]$$

or more generally

$$N_t(p) \text{ iff } (\forall t')[t' \geq t \rightarrow R_t'(P)]$$

The Stoics, on the other hand, dropped the now-relativization of the modalities of possibility and necessity, retaining it only for truth (actuality):

(1) The *actually true* is that which is actually realized *now*

$$T_n(P) \text{ iff } R_n(P)$$

(2) The *possible* is that which is actually realized at *some* (i.e., any) time

$$P(P) \text{ iff } (\exists t)R_t(P)$$

(3) The *necessary* is that which is actually realized at *all* times

$$N(P) \text{ iff } (\forall t)R_t(P)$$

Aristotle's position is in line with that of the Stoics in viewing the necessary as that which is true all of the time,[38] a position faithfully reflected in St. Thomas Aquinas' statement that: *Et sic quidquid semper est, non contingenter semper est, sed ex necessitate.*[39]

In all of this there is no sign of the ramified machinery of temporalized modalities which we find in Arabic texts—and which are unquestionably of Greek provenience. For the roots of this theory we must undoubtedly look to the Stoic doctrine of predication. The Stoics distinguished between three types of qualities:

*Poion* (quality) {
i. *poiotēs* (permanent property)
ii. *schēsis* (enduring state)
iii. *hexis* (transient characteristic)

In construing "quality" (*to poion*) here, we are to work from the top down, and thus have three possibilities:[40]

(1) Only group (i): those qualities that are wholly completed and altogether permanent (*apartizontas kai emmonous ontas*).

[38] J. Hintikka, *op. cit.* Cf. N. Rescher, "Truth and Necessity in Temporal Perspective," in R. M. Gale (ed.), *The Philosophy of Time* (New York, 1967), reprinted in enlarged form in N. Rescher, *Essays in Philosophical Analysis* (Pittsburgh, 1969), pp. 271–302.

[39] *In I De Caelo*, lect. 26, n. 258. And correspondingly: *quod possible est non esse, quandoque non est* (*Summa Theologica*, IA, q. 2, a. 3). Cf. Guy Jalbert, *Necessité et contingence chez saint Thomas d'Aquin et chez ses prédecésseurs* (Ottawa, 1961), pp. 204–206, 224–225, and 228.

[40] I follow E. Zeller, *op. cit.*, pp. 97–99 (especially n. 1 for p. 97); relying also upon Émile Bréhier, *La Théorie des incorporels dans l'ancien Stoicisme* (2nd ed., Paris, 1928), p. 9.

(2) Groups (i) and (ii): not only the permanent qualities (e.g., a man's "being an animal") but the enduring states as well (e.g., "being prudent").
(3) Groups (i)-(iii): adding to (2) also strictly transient qualities (e.g., "walking" or "running").

The distinction between such types of qualities lends itself readily to temporalization in the interpretation of propositions in which they are attributed

A man is an animal *all of the time.*
A prudent man acts wisely *most of the time.*
A healthy man walks *some of the time.*

Distinctions of this sort ultimately derive from Aristotle, who in ch. viii of the *Categories* distinguishes between *states* and *conditions* on the basis that states are more stable and enduring ("is just" vs. "is ill" or again "knows Greek" vs. "is speaking Greek").

One further point is illuminating for the issue of Greek sources. Not only does Avicenna treat the thirteen modalities of al-Qazwīnī, but he classifies them into three groups, as follows:

| *Modes of Necessity* | *Modes of Actuality* | *Modes of Possibility* |
|---|---|---|
| Absolute necessary | Absolute-perpetual | General absolute |
| General conditional | General conventional | General possible |
| Special conditional | Special conventional | Non-necessary existential |
| Temporal | | Non-perpetual existential |
| Spread | | Special possible |

Avicenna's arch-critic, Averroes, rejects this complicated machinery as unwarranted and unnecessary.[41] Averroes himself gives a—to be sure significantly simplified—temporal construction of modalities, presented in a square of opposition as follows:[42]

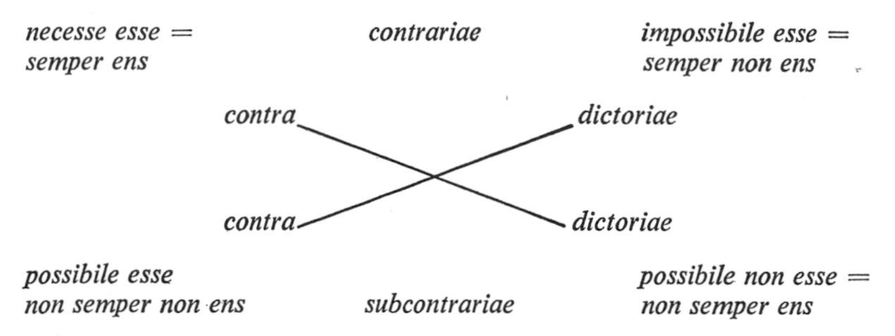


[41] See "Averroes' *Quaesitum* on Assertoric Propositions" in N. Rescher, *Studies in the History of Arabic Logic* (*op. cit.*), pp. 91–105 (see §§ 11–13).
[42] Averroes, *In I De Caelo*, t. 5, f. 85A.

And Averroes affords one interesting historical datum in this connection. According to him, Avicenna maintained in the *Kitāb al-shifā'* that Alexander of Aphrodisias construed necessity to include only the first two of these modes—i.e., the absolute necessary and the general conditional—and that, accordingly he classed the other three modalities of the first group (viz., the special conditional, the temporal, and the spread) as types of actuality. This affords yet another striking datum to the effect that the theory of temporal modalities is of Greek provenience.

It has been known for more than two centuries that there exists in the fine collection of Arabic manuscripts in the Escorial library near Madrid a group of Arabic translations of treatises by Alexander of Aphrodisias, including his refutation of Galen's treatise on possibility. These treatises are reported in Casiri's catalog of 1760,[43] and figure in many later bibliographies.[44] This Galen treatise is not in fact present in the Escorial codex, but the escorial manuscript of Alexander's treatise on Galen on the first mover contains at its outset a single manuscript page (folio 59 *verso*) of Alexander's treatise on Galen on possibility. This one page—the Arabic text of which has been presented elsewhere[45]—consists almost entirely of a single long quotation from Galen's (lost) treatise *De possibilitate*. This is the only portion of this treatise that is known to have survived.[46]

The Galen-fragment at issue is, in fact, of considerable interest for the history of logic. It is well known that Aristotle (and some of the Stoic logicians) proposed to construe the modality of *necessity* in terms of *omnitemporality* (construing, for example, "Water is *necessarily* wet" as "Water is wet *at all times*"). And as we have seen, the

[43] Michael Casiri, *Bibliotheca Arabico-Hispana Escurialensis*, 2 vols. (Madrid, 1760–1770); see vol. I, p. 242, codex no. 794 (the Galen refutation is part of the second part of the second part of the codex, folios 59–69). In the later catalog of H. Derenbourg and H. P. J. Renaud (*Les manuscrits arabes de l'Escurial* [Paris, 1941],) this codex becomes no. 798.

[44] J. G. Wenrich, *De auctorum graecorum versionibus et commentariis*, etc. (Leipzig, 1842), p. 276. L. Leclerc, *Histoire de la medicine arabe*, vol. I (Paris 1876), pp. 216–217; M. Steinschneider, *Die Arabischen Uebersetzungen aus dem Griechischen* (Leipzig, 1893; XII. Beiheft zum *Centralblatt für Bibliothekswesen*), pp. 93–97, and his earlier *Al-Fārābī* (St. Petersbourg, 1869), p. 93.

[45] N. Rescher and M. E. Marmura, *The Refutation by Alexander of Aphrodisias of Galen's Treatise on the Theory of Motion* (Karachi, 1967; Publications of the Central Institute of Islamic Research). The Galen-fragment is translated in N. Rescher, *Temporal Modalities in Arabic Logic* (*op. cit.*).

[46] See the survey of Arabic Alexander MSS in Albert Dietrich, *Die Arabische Version einer unbekannten Schrift des Alexander von Aphrodisias über die Differentia specifia, Nachrichten der Akademie der Wissenschaften in Göttingen*, *Philologisch-historische Klasse* (Jahrgang 1964), no. 2 (see p. 96, item 11). Here the references to the treatises by the Arabic biobibliographers are also listed.

Arabic logicians carried this temporal construction of modality to far greater lengths, elaborating with great sophistication upon the fundamental Aristotelian linkage of modality to time, based on the idea that the necessary is that which happens all of the time.

Now the surprising fact to be gleaned from Galen's report in the fragment here at issue is that *this development of a ramified linkage between time and modality heretofore encountered no earlier than in Arabic logical texts goes back to the earliest Peripatetics*, i.e., to Theophrastus and his followers, who distinguished exactly between rigidly Aristotelian *omnitemporal* necessity (of the "Water is necessarily wet" type described above) and al-Qazwīnī al-Kātibī's "absolute necessity" (of the type: "A man is necessarily an animal" = "A man is an animal so long as he exists").[47] This early linkage of time and modality is a fact of which we have therefore had only somewhat indirect hints, such as that of the following passage from Alexander's *Commentary on "Prior Analytics"*: "For according to him [Aristotle] the necessary is also predicated of the actual (or: *existent*), for what is actually (or: *existentially*) true of something necessarily belongs to (literally: *exists in*) it, so long as it exists. Thus Theophrastus in the first Book of his *Prior Analytics*, speaking of the meanings of the necessary, writes as follows: 'The third meaning is the existent: for when it exists, it is impossible for it not to exist' (fr. 58 Wimmer)" (Ed. M. Wallies, 156: 27–157: 2).

When all these various bits and pieces are put together the result is a fabric of interlocking evidence to the effect that the theory of temporal modalities is of Greek provenience, and that this topic played a role not only in Galen and the Stoics, but also in the Aristotelian tradition from Theophrastus to Alexander of Aphrodisias.

## 11. Temporal Modalities Among the Schoolmen

The Latin medievals appear to have taken the theory of temporal modalities over the Arabs to at any rate a modest extent. In Albert the Great, for example, we find a return to the temporalized modalities in the manner of the Stoics. According to him, modal propositions have a *consignificatio temporis*: a reference to the future is inherent in the modalities *possible* and *contingens*, while the modalities *necessarium* and *impossibile* involve an omnitemporal thesis. The possible is that which will be at some future time; the contingent is that which

[47] For a comparable distinction between absolutely perpetual motion ("of the eternal everlasting things") and non-absolutely perpetual motion ("of things which are not eternal and everlasting but which continue to move as long as they return their nature") see the treatise on the first mover, E68a12–14.

is, but will at some future time not be;[48] the necessary is that which will be always, the impossible never.[49] St. Albert explicitly polemicizes against those who, while granting that the contingent involves reference to the future, deny that the possible does. The possible was already possible prior to its actualization, and so its qualification as possible at this prior juncture must derive from an inherent reference to the future.[50]

In Pseudo-Scotus we find a distinction between four types of temporalized "necessity" (*conditionale* = conditional, *quando* = *as-long-as*, *ut nunc* = as-of-now, and *pro-semper* = for all times).[51] I would conjecture a correspondence with the cognate Arabic ideas along something like the following lines:

| | *Pseudo Scotus* | *Construction* | *al-Qazwīnī* |
|---|---|---|---|
| 1. | *quando* | $\Box\mathscr{E}$ | absolute necessary |
| 2. | *pro semper* | $\forall\mathscr{E}$ | absolute perpetual |
| 3. | *conditionale* | $\Box\mathscr{C}$ | general conditional |
| 4. | *ut nunc*[52] | $\exists\mathscr{E}$ | general absolute |

[48] Or, of course, the negative counterpart of this, viz., that which is not, but will at some future time be. Cf. Boethius' characterization of future contingent propositions as *propositiones, quae cum non sint, eas tamen in futurum evenire possibile est*. This mode of contingency is also present in Albert under the name of *contingens futurum*, and was to give rise to ramified disputes among his successors. See P. Boehner, *The "Tractatus de praedestinatione et de praescientia Dei et de futuribus contingentibus" of William of Ockham* (St. Bonaventure, New York, 1945; Franciscan Institute Publications, No. 2) and L. Baudry, *La Querelle des futurs contingents* (*Louvain* 1465–1475), (Paris, 1950; Etudes de philosophie medievale, fasc. XXXVIII).

[49] *Quatuor enim primi illorum modorum* (*sc. possibile, contingens, necessarium et impossibile*) *compositionem* (*quae consignificat tempus*) *ampliant extra tempus praesens. Possibile enim et contingens ampliant praesens ad futurum, et ad esse, et ad non esse: quia contingens est futurum, et potest esse et non esse. Necessarium autem et impossibile ampliant compositionem ad omne tempus: quia necessarium et impossibile ponunt compositionem in omne tempus: et ideo illi sunt modi speciales facientes totam enuntiationem modalem, necessarium simpliciter omni tempore inesse, et impossibile simpliciter nunquam inesse.* (*Commentaria in libro II, Perihermeneias*, tract II, cap. I; ed. A. Borgnet, vol. I [Paris, 1890], p. 440a, b). On the theory of the temporal "ampliation" of the terms of categorical propositions widely current among the Medieval schoolmen see E. A. Moody, *Truth and Consequence in Medieval Logic* (Amsterdam, 1953).

[50] *Non est verum quod quidam dictum, quod contingens differat a possibili in hoc, quod contingens dicat extensionem temporis in futurum, et possibile non dicat illud: possibile enim ante actum acceptum extenditur in futurum.* (*Ibid.*, tract II, cap. 6; Borgnet, vol. I, p. 452a, b).

[51] I. M. Bochenski, o.p., *Notes historiques sur les propositions modales* (Quebec, 1951), p. 7.

[52] We must here construe "*nunc*" not as *the now* (that is, the now-of-the-present), but as *a now* (that is, some—i.e., *any*—instant). This dual interpretation is standard in the medievals.

If this conjecture—or anything like it—is correct, temporalized modalities also made their way pretty well intact into the Latin scholastics. While this development could have been mediated (or perhaps only strengthened) through Arabic materials, it could also have been indigenous to a purely Latin tradition. Signs of this are already to be found in Boethius:

*Ea vero quae ex necessitate aliquid inesse designat tribus dicitur modis; uno quidem quo ei similis est propositioni quae inesse significat . . . alia vero necessitatis significatio est, cum hoc modo proponimus "hominem necesse est cor habere, dum est atque vivit" . . . alia vero necessitatis significatio est universalis et propria quo absolute praedicat necessitatem . . . possible autem idem tribus dicitur modis; aut enim quod est, possibile esse dicitur . . . aut quod omni tempore contingere potest, cum ea res permanet cui aliquid contingere posse proponitur . . . item possible est quod absolute omni tempore contingere potest . . . ex his igitur apparet alias propositiones esse inesse significantes alias necessarias alias contingentes atque possibiles, quarum necessariarum contingentiumque cum sit trina partitio, singulae ex iisdem partionibus ad eas quae inesse significant referentur; restant igitur duae necessariae et duae contingentes quae cum ea quae inesse significat enumeratae quinque omnes propositionum faciunt differentias; omnium vero harum propositionum aliae sunt affirmativae aliae negarivae.*[53]

Cognate distinctions are also drawn elsewhere. For example, in St. Thomas Aquinas and other medieval scholastics we find a distinction —akin to the modern distinction between logical and physical necessity—between those relationships which are perpetual and (in a sense) necessary: (1) by an eternity that is *a parte ante*, or (2) by an eternity that is *a parte post*. Truths of the former class are necessary by a necessity that turns wholly on the nature of the essences involved —as men are eternally rational and equiangular triangles eternally equilateral. Truths of the second class are necessary by a necessity that devolves from the *au fond* contingent arrangements of this world —as men are eternally mortal, or the northern latitudes eternally cold.[54] Traces of an interest in such temporal modalities are to be found as late as William of Ockham.[55]

[53] Quoted from C. Prantl, *Geschichte der Logik im Abebdlande*, vol. I (Leipzig, 1855; photoreprinted, Graz, 1955), p. 703, n. 150.

[54] See Guy Jalbert, *op. cit.*, pp. 41, 119–120, 137–138, 141–143. This work is primarily concerned with possible and necessary *existents*. A detailed treatment of such existents, primarily in Avicenna, but with some comparisons and contrasts in St. Thomas, can be found in Gerard Smith, "Avicenna and the Possibles," *The New Scholasticism*, vol. 17 (1943), pp. 340–357.

[55] *Summa logicae* (ed. P. Boehner), pt. I, ch. 73, lines 16–49; pt. II, ch's 7, 19–22; pt. III, div. i, ch's 17–19.

## 12. Conclusion

We have come to the end of a long and somewhat complicated account, and a word of retrospective appraisal is in order. Clearly, the Arabic logicians of the Middle Ages—basing their work upon Greek antecedents—were in possession of a complex theory of temporal modal syllogisms, which they elaborated in great and sophisticated detail. When one considers that all reasoning was conducted purely verbally, largely on the basis of somewhat vague examples, without any symbolic apparatus, and even without abbreviative devices, one cannot but admire the level of complexity and accuracy. The intricacy of the modal notions at issue was such that, in the context of complex interrelationships—and especially in the modal syllogistic—errors tended to creep in. Nevertheless, the basic ideas were clear and the guiding intuitions of their elaboration straightforwardly intelligible. It is, I believe, safe to say that—barring possible new findings from the side of medieval Latin scholasticism—the logical theory of temporal concepts was carried to a higher point in Arabic logic than at any subsequent juncture prior to our own times.

The logical acumen of these medieval scholars was of a very high order indeed. But their successors were not able to maintain this standard. Sprenger remarks in his translation of the *Shamsiyyah* of Qazwīnī:

> [The paragraphs dealing with modalized inferences] are omitted in the translation because they contain details on modals which are of no interest. The last named four paragraphs are also omitted in most Arabic text books on Logic, and are not studied in Mohammedan Schools.[56]

When the logical tradition of Islam passed from the hands of the scholars into that of the schoolmasters, the standard of work went into a not surprising decline. The medievals had a firmer grasp.[57]

[56] A. Sprenger, *op. cit.*, p. 25.

[57] This essay is a revised version of a paper originally written in collaboration with Arnold vander Nat and published under the title "The Arabic Theory of Temporal Modal Syllogistic" in G. Hourani (ed.), *Essays in Islamic Philosophy and Science* (Albany, 1973).

III.

# Leibniz and the Evaluation of Possible Worlds

## 1. STAGESETTING

PRESENT-DAY studies of Leibniz have remarkably little to say about one altogether central doctrine of his metaphysical system —his theory of the *standard of value* determinative of the relative goodness of possible worlds. Leibniz's manifold discussion of the metaphysical perfection of worlds are given short shrift, presumably being viewed by most recent commentators as something of a vestigial remnant of scholastic sophistry. This, to my mind, is unfortunate, for it seems to me that his teachings on this topic are among the most seminal and significant aspects of Leibniz's entire system.

Which of the alternative possible worlds he contemplates *sub ratione possibilitatis* is God to select for actualization? This question clearly poses one of the central issues of Leibniz's philosophy. He bitterly opposed the position of Descartes and Spinoza, whom Leibniz took to maintain the indifference and arbitrariness of God's will. Leibniz again and again insisted that there is an independent standard of the perfection of things—a standard determined by considerations of objective necessity, which the preferences and decisions of the deity could alter no more than the sum of two plus two. Moreover, this standard is not to apply simply to the actual domain of the real world, but is operative throughout the modally variant sphere of the merely possible as well. Leibniz held that possibilities are objectively good or bad by a standing "*règle de bonté*" that operates altogether independently of the nature of existence and of the will of God. Indeed this standard of goodness is the basis for two crucial modal distinctions: (1) since God acts for the best in all of his actions, that of creation pre-eminently included, it serves to demarcate the actual from the possible, and (2) given God's goodness, it renders the sphere of the actual as *necessary*—not in the absolute or metaphysical sense of this term, but in its relative or moral sense. The standard of goodness is accordingly pivotal for the operation of modal distinctions in the system of Leibniz.

But what is this criterion which a God who seeks to actualize *the best* of possible worlds employs in identifying it? By what criterion of merits does God determine whether one possible world is more or less perfect than another? This standard, Leibniz maintains, is the combination of *variety* and *order*.

Accordingly the best of possible worlds is that which successfully manages to achieve the greatest richness of phenomena (*richesse des effects* [*Discours de métaphysique*, §5], *fecondité* [*Théodicée*, §208] *varietas formarum* [*Phil. Werke*, ed. Gerhardt, VII, §303]) that can be combined with (*est en balance avec* [*Discours de métaphysique*, §5], *sont les plus fécondes par rapport à* [*Théodicée*, §208]) the greatest simplicity of laws (*la simplicité des voyes* [*Discours de métaphysique*, §5 and *Théodicée*, §208], *le plus grand ordre* [*Principles of Nature and of Grace*, §10]).

In the elegant essay *De rerum originatione radicali* Leibniz puts the matter as follows:

> Hence it is most clearly understood that among the infinite combinations of possibles and possible series, that one actually exists by which the most of essence or of possibility is brought into existence. And indeed there is always in things a principle of determination which is based on consideration of maximum and minimum, such that the greatest effect is obtained with the least, so to speak, expenditure. And here the time, place, or in a word, the receptivity or capacity of the world may be considered as the expenditure or the ground upon which the world can be most easily built, whereas the varieties of forms correspond to the commodiousness of the edifice and the multiplicity and elegance of its chambers. (Tr. L. E. Loemker.)

It is worthwhile to consider in some detail the individual components of Leibniz's two-factor criterion of variety and richness of phenomena on the one hand and lawfulness or order on the other.

The reference to lawfulness clearly carries back straightaway to Greek ideas (balance, harmony, proportion). The origination of cosmic order is a key theme in the Presocratics (e.g., Anaximander) and becomes one of the great central issues of ancient philosophy with the writing of Plato's *Timaeus*. Afterwards, it of course plays a highly prominent role in the church fathers and achieves a central place in scholasticism in connection with the Cosmological Argument for the existence of God.

The prime factor in Leibniz's theory is not, however, lawfulness as such, but the simplicity or economy of laws. Leibniz, as we know, held that *every* possible world is lawful. As he puts it in §6 of the *Discours de métaphysique*: *on peut dire que, le quelque manière que Dieu auroit créé le monde; il auroit toujours esté regulier et dans un certain ordre general.* The critical difference between possible worlds in point of lawfulness is thus not whether there are laws or not—there *always* are —but whether these laws are relatively simple.

Regrettably, Leibniz nowhere treats in detail the range of issues involved in determining the relative simplicity of bodies of laws, and indeed he does not seem to be fully aware of the complexities that inhere in the concept of simplicity. Perhaps he did not think it necessary to go into details because it is, after all, God and not us imperfect humans by whom this determination is to be made. And in any case, it is clear enough in general terms what he had in mind. No one for whom the development of classical physics in its "Newtonian" formulation as a replacement of Ptolemaic epicycles and Copernican complexity was a living memory could fail to have some understanding of the issues. (It might be observed parenthetically that Leibniz would surely have viewed with approval and encouragement the efforts by N. Goodman, J. G. Kemeny, and others during the 1950's[1] to develop an exact analysis of the concept of simplicity operative in the context of scientific theories.) So much then for simplicity of laws; let us turn to variety.

The situation as regards *variety* is even somewhat more complicated. As Leibniz considers it, variety has two principal aspects: fullness or completeness or comprehensiveness of content on the one hand, and diversity and richness and variation upon the other. All these factors are certainly found in ancient writers. They are notable in the *Timaeus*[2] and play a significant role in Plotinus and neoplatonism.[3] The church fathers also stressed the role of completeness and fullness (*fecunditas*) as a perfection, and it is prominent in St. Thomas's treatment of the cosmological argument and also in the later schoolmen. The recognition of the metaphysical importance of variety is thus an ancient and stable aspect of Platonic tradition.

But a new, Renaissance element is present in Leibniz's treatment of this theme: the aspect of *infinitude* that did not altogether appeal to the sense of tidiness of the more fastidious Greco-Roman mentality. The Renaissance evolution of a spatially infinite universe from the finite cosmos of Aristotle is a thoroughly familiar theme. And it

[1] Nelson Goodman, *The Structure of Appearance* (Cambridge, Mass., 1951; 2nd ed., Indianapolis, 1966); *idem.*, "Safety, Strength, Simplicity," *Philosophy of Science*, vol. 28 (1961), pp. 150–151; John G. Kemeny, "The Use of Simplicity in Induction," *The Philosophical Review*, vol. (1953), pp. 391–408.

[2] *Timaeus* 33B; cf. F. M. Cornford's and T. L. Heath's comments *ad loc.*

[3] Recall the stress on generative energy and creative power in the *Enneads* of Plotinus, and passages like "This earth of ours is full of varied life-forms and of immortal beings, to the very heavens it is crowded" (*Enn.* II, 8; McKenna).

represents a development that evoked strong reactions. One may recall Giordano Bruno's near-demonic delight with the break-up of the closed Aristotelian world into one opening into an infinite universe spread throughout endless spaces. Others were not delighted but appalled—as, e.g., Pascal was frightened by "the eternal silence of infinite spaces" of which he speaks so movingly in the *Pensées* (§§205–206). An analogous development occurred with respect to the strictly *qualitative* aspects of the universe. Enterprising and ardent spirits like Paracelsus, Helmont, and Bacon delighted in stressing a degree of complexity and diversity not envisaged in the ancient authorities. A vivid illustration of this welcoming of diversity is Leibniz's insistence that the variety of the world is not just a matter of the number of its substances, but of the infinite multiplicity of the forms or kinds they exemplify. He would not countenance a *vacuum formarum:* but taught that infinite gradations of kind connect any two natural species. Accordingly he was positively enthusiastic—as no classically fastidious thinker could have been—with the discovery by the early microscopists of a vast multitude of little squirming things in nature. Leibniz's concern for qualitative infinitude as an aspect of variety represents a distinctly modern variation of an ancient theme. In giving not only positive but even paramount value to variety, complexity, richness and comprehensiveness, Leibniz expresses with characteristic genius, the Faustian outlook of modern European man.

His concern for an objective standard of cosmic valuation gives Leibniz a central place within the tradition of *evaluative metaphysics*. The paternity of this branch of philosophy may unhesitatingly be laid at Plato's door, but as a conscious and deliberate philosophical method it can be ascribed to Aristotle, whose preoccupation in the *Metaphysics* with the ranking schematism of prior/posterior—for which see especially chap. 11 of Bk. 5 (Delta), and chap. 8 of Bk. 9 (Theta)—is indicative of his far-reaching concern with the evaluative dimension of metaphysical inquiry.[4]

Now as has been indicated, there is nothing new and original in a stress upon variety and order as aspects of the perfection of creation. Both of these factors are prominent in the philosophical atmosphere of the late 17th century. Thus for example in §17 of Bk. I of Malebranche's *Traité de la Nature et de la Grace* (published in 1680) we read:

[4] For a more comprehensive treatment of this evaluative dimension of metaphysics see pp. 230–243 of N. Rescher, *Essays in Philosophical Analysis* (Pittsburgh, 1969).

> Or ces . . . lois sont si simples, si naturelles, & en même temps si fécondes que quand on n'auroit point d'autres raisons pour juger que ce sont établies par celui qui agit toujours par les voies les plus simples, dans l'action duquel il n'y a rien qui ne soit réglé, & qui la proportionne si sagement avec son ouvrage, qu'il opère une infinité de merveilles par un trés-petit nombre de volontez.[5]

What can Leibniz add to this very Leibnizian passage?

What he adds is exactly to establish these two long-prominent *aspects* of the world's perfection as jointly operative and mutually conditioning criteria joined within *a single two-factor standard of the perfection* of a possible world. What is specifically characteristic of Leibniz is the idea of combination and balance of these factors in a state of mutual tension.

But why should it be plausible to take this step and establish variety and order as conjoint but potentially conflicting yardsticks of perfection? The basis of plausibility of this Leibnizian standard rests upon a whole network of analogies, three of which are clearly primary:

(1) *Art.* Throughout the fine arts an excellent production requires that a variety of effects be combined within a structural unity of workmanship. Think here of the paradigm of Baroque music and architecture.

(2) *Statecraft.* Excellence can only be achieved in the political organization of affairs when variety (freedom) is duly combined with lawfulness (= order and the rule of law).

(3) *Science.* Any really adequate mechanism of scientific explanation must succeed in combining a wide variety of phenomena (fall of apple, tides, moon) within the unifying range of a simple structure of laws (gravitation).

All of these diverse paradigms meet and run together in Leibniz's thinking. Like lesser luminaries such as Spengler and Ernst Cassirer, Leibniz had an extraordinarily keen eye for the perception of deep subsurface analogies. A true systematizer, he likes to exploit vast overarching connections and exhibits an extraordinary talent for transmitting a discernment of common structures into the formulation of an illuminating and fruitful theory.

## 2. Mathematico-Physical Inspiration

One immediately striking feature of the Leibnizian standard of metaphysical perfection in terms of orderliness and variety is that this

[5] Ed. G. Dreyfus (Paris, 1958), p. 187.

is a *conflict-admitting two-factor criterion*, and as such contrasts sharply with the long series of monolithic *summum bonum* theories that have so generally been in vogue in ethics—both before the time of Leibniz and afterwards, down to our own day.

From this aspect, Leibniz's position is strikingly reminiscent of that taken by his philosophical idol, Plato, in the *Philebus*. Plato there holds, as against various simplistic philosophical doctrines, that the good life cannot rest on (say) knowledge or beauty or pleasure alone, but requires a *mixture* of such factors in which each constituent has its proper share. But, of course, most of the subsequent ethical tradition from the Stoics and Epicureans to the utilitarians and Freudians were impatient with Platonic complexity and have been eager to press the ethical predominance and primacy of one unique monolithic factor (pleasure, the good will, personal adjustment, or whatever). Moreover, the various Platonic ingredients themselves (say pleasure and knowledge) are not in any obvious way in conflict with each other, since—as Plato himself stresses—there is the "pure" pleasure that we can take in knowdge as such. Leibniz's handling of this problem of mixture is, however, substantially more sophisticated than Plato's, because Leibniz, unlike Plato, does not envisage a resolution in terms of a *fixed proportion*, but rather one in terms of the sort of dynamic tension that mathematical interrelationships of a more complicated (nonlinear) type make possible. Put in economic terminology, Leibniz thinks of these factors as related not by a fixed exchange ratio, but by variable trade-offs with diminishing marginal returns for both parameters.

The immediately striking feature of the criterion is that the two operative factors are *opposed* to one another and pull in opposite directions. On the one hand, a world whose only metal is (say) copper, or whose only form of animal life is the amoeba, will obviously have a simpler structure of laws because of this impoverishment. On the other hand, a world whose laws are more complex than the rules of the astrologers demands a wider variety of occurrences for their exemplification. Clearly the less variety a world contains—the more monotonous and homogeneous it is—the simpler its laws will be; and the more complex its laws, the greater the variety of its phenomena must be to realize them. Too simple laws produce monotony; too varied phenomena produce chaos. So these two criterial factors of order and variety are by no means cooperative, but stand in a relationship of mutual tension and opposition.

Leibniz's conception of the deity's way of proceeding in the selection of one of the possible worlds for actualization can be represented and illustrated by the sort of infinite-comparison process familiar from the calculus and the calculus of variations. In his interesting and very important little essay *Tentamen Anagogicum* Leibniz puts the matter as follows:

> The principles of mechanics themselves cannot be explained geometrically, since they depend on more sublime principles which show the wisdom of the Author in the order and perfection of his work. The most beautiful thing about this view seems to me that the principle of perfection is not limited to the general but descends also to the particulars of things and of phenomena and that in this respect it closely resembles the method of *optimal forms*, that is to say, of forms *which provide a maximum or minimum*, as the case may be—a method which I have introduced into geometry in addition to the ancient method of *maximal and minimal quantities*. For in these forms or figures the *optimum* is found not only in the whole but also in each part, and it would not even suffice in the whole without this. For example, if in the case of the curve of shortest descent between two given points, we choose any two points on this curve at will, the part of the line intercepted between them is also necessarily the line of shortest descent with regard to them. It is in this way that the smallest parts of the universe are ruled in accordance with the order of greatest perfection.[6]

In taking as measure of perfection the combination of two essentially conflicting factors, Leibniz unquestionably drew his inspiration once again from mathematics—as he so often does. Determining the maximum or minimum of that surface-defining equation which represents a function of two real variables specifically requires those problem-solving devices for which the mechanisms of the differential calculus were specifically devised. Unlike the relative mathematical naivete of the old-line, monolithic, single-factor criterion, the Leibnizian standard of a plurality of factors in nonlinear combination demands the sort of mathematical sophistication that was second nature to him.

## 3. Ethical Ramifications

I turn now to the ethical aspect of Leibniz's conception of metaphysical perfection. The welfare of the *spirits*—the highest grade of monads that comprise the "City of God"—is, according to Leibniz, a fundamentally primary consideration in the deity's comparative evaluation of alternative worlds: "There is no room for doubt that the felicity of the spirits is the principal aim of God and that He puts this purpose into execution as far as the general harmony will permit."[7] This second *moral* approach, of course, leads to another

[6] Tr. in L. E. Loemker (ed.), *Gottfried Wilhelm Leibniz: Philosophical Papers and Letters* (Dordrecht, 1969; 2nd ed.), p. 478.

[7] *Monadology*, §86. Cf. *Principles of Nature and of Grace*, §§14–15.

and seemingly quite different *metaphysical* criterion of merit in the assessment of possible worlds. It points, namely, not to an *ontological* beneficence that interests itself in such substantially universal and abstract considerations as rich phenomena subject to simple laws, but to the specifically *ethical* beneficence of a creator seeking preeminently to assure the welfare of the spirits.

Now the metaphysical concept of goodness in terms of variety and order seemingly stands in potential conflict with that emphatically moral criterion inherent in acknowledging God's primary responsibility to the spirits. To all appearances, its adoption seemingly marks Leibniz as a tough-minded system-builder, in whom the cold-blooded metaphysician triumphs over the moral philosopher and the Christian theologian.

Within the framework of Leibnizian commitments the only really satisfactory way to validate the step from God's metaphysical concern for the ontological perfection of the world to God's ethical concern for the moral welfare of spirits is by demonstrating that the metaphysical features of lawfulness and variety in the world are *preconditions for the welfare of spirits*. And the general lines along which such a demonstration can be given is readily imagined. After all, every possible substance embraces a multiplicity (i.e., variety) within a unity of lawful development (i.e., order). And spirits as the most fully and perfectly developed substances must be presumed such that their welfare and felicity lies in realizing these conditions to an extremely high degree. Moreover, since Leibnizian substances are extremely environment-sensitive (that is perceive and reflect to a high degree the condition of the other substances that constitute their world) they cannot achieve the requisite *inner* complexity and order unless this is externally present in their cosmic surroundings. (Think here of Leibniz's favorite dictum, *sympnoia panta*.)

In short, it would seem that the political analogy is determinative for Leibniz. His paradigm is the view that a person can realize the full development of his personality essential for maximal achievement of his own welfare only in a social life-environment that affords a proper combination of law and order on the one hand and on the other the variety of stimuli and resources requisite for the full and free development of one's potential and fulfillment of one's basic needs as a person. Human development in full requires the appropriate sort of social environment. And analogously, monadic development in highest degree (= maximizing the welfare of spirits) requires a corresponding benign cosmic environment. On both sides, a combination of orderliness with diversity is required.

It is worth noting that the fundamentality of the political analogy manifests itself not only on the side of the creation, but on that of the creator as well. For he is not only the source of all being but also its ruler, and in particular stands assumes the special role of reigning prince over the community of spirits, a concept that leads straightway to the central Leibnizian theory of the "City of God."

## 4. Epistemological Implications

I come now to the principal points of this discussion, the transformation of the preceding metaphysical considerations into epistemological ones.

In dealing with the metaphysical standard of perfection and issues of the ethics of creation we are, of course, looking at things from a God's-eye point of view. At that level one is involved with the God-oriented issue of which among innumerable possible worlds is to be realized. Accordingly, one's concern is with the *ontological* issue of what is to be real among diverse alternatives. The standard of variety-cum-orderliness is to be considered in the light of this metaphysical issue.

But in turning from metaphysics to epistemology we leave this God's-eye perspective behind. Our concern now is not with the God-oriented issue of which existential possibility is to be real (i.e., realized) but with the very human problem of which among the many possibilities that seem so to us are actually real. The problem now is not which ontological possibility is to be actualized but which epistemological possibility is to be recognized as actual—i.e., veridical.

Leibniz treats this epistemological issue in one of his most powerfully seminal works, the little tract *De modo distinguendi phaenomena realia ab imaginariis*. How does the golden mountain I imagine differ from the real earthen, rocky, and wooded mountain I see yonder? Primarily in two respects: internal detail and general conformity to the course of nature. Regarding the internal detail of vividness and complexity Leibniz says:

> We conclude it from the phenomenon itself if it is vivid, complex, and internally coherent [*congruum*]. It will be vivid if its qualities, such as light, color, and warmth, appear intense enough. It will be complex if these qualities are varied and support us in undertaking many experiments and new observations; for example, if we experience in a phenomenon not merely colors but also sounds, odors, and qualities of taste and touch, and this both in the phenomenon as a whole and in its various parts which we can further treat according to causes. Such a long chain of observations is usually begun by design and selectively

> and usually occurs neither in dreams nor in those imaginings which memory or fantasy present, in which the image is mostly vague and disappears while we are examining it.[8]

Regarding the second aspect of coherence, Leibniz says:

> A phenomenon will be coherent when it consists of many phenomena, for which a reason can be given either within themselves or by some sufficiently simply hypothesis common to them; next, it is coherent if it conforms to the customary nature of other phenomena which have repeatedly occurred to us, so that its parts have the same position, order, and outcome in relation to the phenomenon which similar phenomena have had. Otherwise phenomena will be suspect, for if we were to see men moving through the air astride the hippogryphs of Ariostus, it would, I believe, make us uncertain whether we were dreaming or awake.[9]

He elaborates this criterion in considerable detail:

> But this criterion can be referred back to another general class of tests drawn from preceding phenomena. The present phenomenon must be coherent with these if, namely, it preserves the same consistency or if a reason can be supplied for it from preceding phenomena or if all together are coherent with the same hypothesis, as if with a common cause. But certainly a most valid criterion is a consensus with the whole sequence of life, especially if many others affirm the same thing to be coherent with their phenomena also, for it is not only probable but certain, as I will show directly, that other substances exist which are similar to us. Yet the most powerful criterion of the reality of phenomena, sufficient even by itself, is success in predicting future phenomena from past and present ones, whether that prediction is based upon a reason, upon a hypothesis that was previously successful, or upon the customary consistency of things as observed previously.[10]

Thus Leibniz lays down two fundamental criteria for the distinguishing of real from imaginary phenomena; the vividness and complexity of inner detail on the one hand and the coherence and lawfulness of mutual relationship upon the other.

Now the interesting and striking fact about this sector of Leibnizian epistemology is *its complete parallelism* with his ethical metaphysics of creation. In both cases alike, the operative criterion of the real resides in a combination of variety and orderliness. This is certainly no accident. One cannot but sense the deep connection at work here. Let us attempt to illuminate it.

Leibniz's line of thought begins with a theologico-metaphysical application of ethical theory: the doctrine that God will chose for

[8] Tr. L. E. Loemker, *op. cit.*, pp. 363–364.
[9] Tr. L. E. Loemker, *op. cit.*, p. 364.
[10] *Ibid.*

actualization that one among all possible worlds which qualifies as "the best." The implementation of this doctrine, of course, calls for a *metaphysical standard of relative perfection*, a requirement filled by the Leibnizian criterion of lawfulness and variety.

Given this starting-point it is natural to invoke the logical principle of *adaequatio intellectu ad rem* to the effect that, as Spinoza puts it, "the order and connection of ideas is the same as the order and connection of things." Appeal to this principle serves to transmute our metaphysical standard of perfection as used ontologically for bridging the metaphysical division between *possibility and reality* into an epistemological standard for bridging the division between *appearance and reality*. In this way, Leibniz shifts the application of the fundamental criterion of variety and orderliness from God's realization-selection of a real among possible worlds to man's recognition-selection of a real among apparent phenomena.

Leibniz's line of thought thus in effect exhibits the striking feature of using a logical doctrine, the correspondence theory of truth and reality (*adaequatio intellectu ad rem*), to validate an epistemological coherence theory of truth and reality in terms of the ethico-metaphysical standard of perfection that he views as operative in God's creation choice. This complex and fruitful conjoining of different elements is altogether typical of Leibniz's ingenuity as a philosophic system-builder.

## 5. Leibniz as Pioneer of the Coherence Theory of Truth

The coherence theory of truth has played a central role in thinking of the Anglo-American idealists from Bradley, Bosanquet, and Joachim to A. C. Ewing and Brand Blanshard in our own day. Moreover this theory of truth has had a definite appeal for some members of the Vienna school of logical positivism (O. Neurath, R. Carnap [at one brief stage], C. G. Hempel [in some passages]). There is no time here to go into details, and I shall simply presuppose familiarity with the theory and its development.[11] But it is relevant to our present concerns to note that the modern coherence theorists articulate a criterion of truth that revolves around exactly the two Leibnizian factors of variety and order.

To establish this point, I shall confine myself to citing the best-known idealist exponent of the coherence theory, the English metaphysician F. H. Bradley, who writes as follows:

[11] For a fuller discussion, including references, see N. Rescher, *The Coherence Theory of Truth* (Oxford, 1973).

> There is a misunderstanding against which the reader must be warned most emphatically. The test which I advocate is the idea of a whole of knowledge as wide and as consistent as may be. In speaking of system I mean always the union of these two aspects, and this is the sense and the only sense in which I am defending coherence. If we separate coherence from what Prof. Stout calls comprehensiveness, then I agree that neither of these aspects of system will work by itself. How they are connected, and whether in the end we have one principle or two, is of course a difficult question. . . . All that I can do here is to point out that both of the above aspects are for me inseparably included in the idea of system, and that coherence apart from comprehensiveness is not for me the test of truth or reality.[12]

Bradley thus insists emphatically upon conjoining in his own coherence criterion of truth exactly the two Leibnizian factors of order (= coherence) and variety (= comprehensiveness).

These very brief indications should, I think, suffice to show that Leibniz must be viewed as a pioneer of this line of thought and he beyond question qualifies as one of the fathers of the coherence theory of truth.

There is, to be sure, a crucial difference between Leibniz and the English neo-Hegelians who espoused the coherence theory of truth at the end of the last century. They are separated by the vast gulf of Kant's Copernican Revolution.

Unlike Leibniz, the modern idealists usually abandoned altogether the traditional correspondence-to-fact idea of truth, and looked upon coherence as affording not an epistemological *criterion* of truth but a logical *definition* of it. They gave up as useless baggage the whole idea of correspondence with an *an sich* reality. This, of course, is a position which Leibniz was unable to take, so that, naturally enough, he remained in the pre-Kantian dogmatic era in which the conception of truth as agreement with an altogether extra-mental reality was inevitable.

Nevertheless, though on the metaphysical side Leibniz stays with the dogmatists in his acceptance of an *an sich* reality as the ultimate metaphysical basis of truth, still, on the *epistemological* side, he very definitely foreshadows Kant. Consider the following passage, again from the important little essay *De modo distinguendi phaenomena realia ab imaginariis:*

> We must admit it to be true that the criteria for real phenomena thus far offered, even when taken together, are not demonstrative, even though they have the greatest probability; or to speak popularly, that

[12] "On Truth and Coherence" in *Essays on Truth and Reality* (Oxford, 1914), pp. 202–218 (see pp. 202–203).

they provide a moral certainty but do not establish a metaphysical certainty, so that to affirm the contrary would involve a contradiction. Thus by no argument can it be demonstrated absolutely that bodies exist, nor is there anything to prevent certain well-ordered dreams from being the objects of our mind, which we judge to be true and which, because of their accord with each other, are equivalent to truth so far as practice is concerned. Nor is the argument which is popularly offered, that this makes God a deceiver, of great importance. . . . For what if our nature happened to be incapable of real phenomena? Then indeed God ought not so much to be blamed as to be thanked, for since these phenomena could not be real, God would, by causing them at least to be in [mutual] agreement, be providing us with something equally as valuable in all the practice of life as would be real phenomena. What if this whole short life, indeed, were only some long dream and we should awake at death, as the Platonists seem to think? . . . Indeed, even if this whole life were said to be only a dream, and the visible world only a phantasm, I should call this dream or this phantasm real enough if we were never deceived by it when we make good use of reason. But just as we know from these marks which phenomena should be seen as real, so we also conclude, on the contrary, that any phenomena which conflict with those that we judge to be real, and likewise those whose fallacy we can understand from their causes, are merely apparent.[13]

The lesson of this passage is clear. In *epistemology*, at any rate, we have no need for the correspondentist conception of truth as *adaequatio ad rem*. The distinction between appearance and reality is indeed crucial, but it is *for us* a distinction strictly to be drawn *wholly within* the domain of phenomenal reality, and is not a distinction between phenomenal reality on the one hand and noumenal reality on the other.

Thus, while from the God's-eye perspective of his metaphysics Leibniz remains the author of the "system of the Monadology," from the man's-eye perspective of epistemology Leibniz is very much the colleague not only of Kant himself but also of the latter-day idealistic advocates of the coherence theory of truth.

[13] Tr. L. E. Loemker, *op. cit.*, pp. 364–365.

IV.

# Kant and the "Special Constitution" of Man's Mind

## (The ultimately Factual Basis of the Necessity and Universality of A Priori Synthetic Truths in Kant's Critical Philosophy)

### 1. The Thesis

I SHALL maintain that in Kant's system the fundamental principles of the theory of cognition ("theoretical reason"), the theory of evaluation ("judgment"), and of the theory of action ("practical reason") rest on an ultimately *factual* foundation, to wit, "the special constitution" peculiarly characteristic of the human mind. Accordingly, we must recognize that the universality and necessity of synthetic *a priori* propositions as established by Kant's line of critical argumentation are not absolute (or categorical) but relative (or hypothetical): they are specifically relativized to the workings of the *human* intellect, the peculiarly characteristic structure or "special constitution" of *our* cognitive faculties. In the areas of theoretical reason, evaluative judgment, and practical reason, we thus have to do, in a Kantian context, not with what is universal or necessary for *any* rational creature whatsoever, but merely what is relatively necessary for rational creatures equipped with a mind of a type akin to our own.

It becomes essential for clarity to apply in this Kantian context the distinction between thing-pertaining (*de re*) and thesis-pertaining (*de dicto*) universality and necessity. Consider a proposition like "All (plane) triangles have angles summing to 180°." From the perspective of what is internal to the judgment itself, and manifestly explicit in it, universality and necessity obtain unrestrictedly: "*All* (plane) triangles whatsoever *must of necessity* have angles summing to 180°." By contrast, universality and necessity can also arise from a judgment-external perspective, namely with respect to the range of subjects (intelligences) *for* which (rather than the range of objects *to* which) this judgment applies. Now it is merely this "external," *de dicto* range of the universality and necessity of our judgments that is specifically limited (viz., to us humans). And it follows from this that the "internal" *de re* range of the judgment must accordingly be understood as also tacitly or implicity delimited: "All (plane) triangles, *as we humans do AND CAN conceive them*, will necessarily have angles summing to 180°."

It is in *this* sense of a tacit, covert, implicit limitation in the universality and necessity of our judgments to the specifically human context that I wish to argue for the ultimately factual basis of necessity and universality in Kant's Critical Philosophy.

## 2. Forms of Intuition and Categories of Understanding

There is no need to go to great lengths to establish Kant's acknowledgment that the forms of intuition at issue in the Transcendental Aesthetic are forms of *human* intuition, relating to the perceptual experience enjoyed by us mortal men. He could scarcely be more explicit on this point:

> But intuition takes place only in so far as the object is given *to us*. This again is possible, *for us men at any rate* (*uns Menschen wenigstens*) in so far as the mind is affected in a certain way. The capacity (receptivity) for receiving representations through the mode in which *we* are affected by objects, is called "sensibility." Objects are given *to us* by means of sensibility, and it alone yields *us* intuitions. . . . But all thought must, directly or indirectly . . . relate ultimately to intuitions, and therefore, *with us*, to sensibility, because in no other way can an object be given *to us*. (A19=B33, my italics.)[1]

Thus it is clear that on Kant's view the way in which objects are "given" in perceptual experience is relativized to the particular features characterizing our specifically human faculty of intuition. "*Our nature is so constituted* that *our* intuition can never be other than sensible; that is, it contains only the mode in which *we* are affected by objects" (A51=B75, my italics).

That other sorts of intelligent creatures (e.g., such higher intelligences as angels or gods) would inevitably have to intuit objects in the forms of spatiality and temporality characteristic of us men is something that Kant would expressly deny. Of space and time Kant says explicitly that "they belong only to the form of intuition, *and therefore to the subjective constitution of our mind*, apart from which they could not be ascribed to anything whatsoever" (A23=B38, my

[1] I cite Kant's works in the following translations: *Critique of Pure Reason* (=C.P.R.) tr. N. K. Smith (New York, 1965); *Critique of Practical Reason* (=C.Pr.R.) tr. L. W. Beck (New York, 1958); *Critique of Judgment* (=C.J.) tr. J. H. Bernard (New York, 1951); *Foundations of the Metaphysic of Morals* (=F.M.M.) tr. L. W. Beck (New York, 1959); *Prolegomena to Any Future Metaphysics* (=Prol.) tr. L. W. Beck (New York, 1950).

italics). Thus space is "the subjective condition of [our] sensibility, under which alone outer intuition is possible *for us*" (A26=B43, my italics), so that "it is, therefore, *solely from the human standpoint* that we can speak of space" (A26=B42). And the same holds of time, which "is nothing but the subjective condition under which alone intuition can take place *in us*" (A37=B49, my italics). Accordingly "if we abstract from *our* mode of inwardly intuiting ourselves . . . and so take objects as they may be in themselves, then time is nothing" (A34=B51, Kant's italics), so that "time is *a purely subjective condition of our human intuition*" (A35=B51, my italics).

Thus space and time relate specifically to *human* sensibility:

> "What objects may be . . . apart from all this receptivity of *our* sensibility, remains completely unknown to us. We know nothing but our mode of perceiving them [i.e., objects]—*a mode which is peculiar to us, and not necessarily shared in by every being, though, certainly, by every human being*" (A42=B59; my italics).

Of course, it is possible that beings other than men also possess a sensibility that is subject to these same conditions, but this is something of which in principle we can have no information:

> This mode of intuiting in space and time need not be limited to human sensibility. It may be that all finite, thinking beings necessarily agree with man in this respect, although we are not in a position to judge whether this is actually so. (B72).

Thus the forms of our sensibility (viz., space and time) are of specifically anthropoid bearing. And accordingly, the cognitive disciplines that fall within their orbit (viz., geometry, arithmetic) embrace propositions whose necessity and universality is specifically man-relative.

The situation as regards the formal concepts of "the understanding" parallels that of the forms of "the sensibility"—namely it is specifically *our human understanding* that is at issue. Even as the forms of sensibility represent *subjective conditions of* [*human*] *intuition*, so the categories represent *subjective conditions of* [*human*] *thought* (A89=B122). For all objects of thought must "conform to the conditions which the [human] understanding requires for the synthetic unity of thought" (A90=B123). That "we cannot think an object save through [the] categories" (B165) is clearly as much a fact about *us* as about *them*. The categories must be recognized as "*a priori* conditions of the possibility of [human] experience" (A94=B126), and their validation (in the "Deduction of the Categories") compels us "to penetrate . . . deeply into the first grounds of the possibility of *our* knowledge" (A98, my italics).

The basis of this deduction, which Kant denominates as "the transcendental unity of apperception," lies in the principle that any cognition can be accompanied by the reflexive indication of self-consciousness, the "I think".

> This principle is not, however, to be taken as applying to every possible understanding, but only to that understanding through whose pure apperception, in the representation "I am", nothing manifold is given. An understanding which through its self-consciousness could supply to itself the manifold of intuition—an understanding, that is to say, through whose representation the objects of the representation should at the same time exist—would not require, for the unity of consciousness, a special act of synthesis of the manifold. For the human understanding, however, which thinks only, and does not intuit, that act is necessary. It is indeed the first principle of the human understanding, and is so indispensable to it that we cannot form the least conception of any other possible understanding, either of such as is itself intuitive or of any that may possess an underlying mode of sensible intuition which is different in kind from that in space and time. (B139; cf. B153.)

Since all else in the legitimation of the categories revolves around this point, it is clear that here again the ultimate source of universality and necessity is again seen to reside in the specific constitution of the human mind.

The specific constitutive features of the human mind manifested on the side of intuition by two forms of sensibility and on that of the understanding by the twelve categories must be accepted as given ultimates incapable of explanation or rationalization of any sort:

> This peculiarity of our understanding, that it can produce *a priori* unity of apperception solely by means of the categories, and only by such and so many, is as little capable of further explanation as why we have just these and no other functions of judgment, or why space and time are the only forms of our possible intuition (B145–146).

Every philosophical system is bound to come up against the realities of surd fact somewhere along the line. With Kant we reach it at the level of the cognitive constitution of the human mind. In fact, this relativization of the *a priori* facets of our knowledge to the basic *modus operandi* of the human intellect is exactly what Kant's Copernican Revolution is all about. For at just this point, a reference to "the special constitution" of our human faculties of intuition and of understanding becomes crucial. (Compare Bxvii-xviii.) In the *Critique of Judgment*, Kant formulates this key point with naked bluntness and almost brutal force:

> If we look merely to the way in which anything can be *for us* (according to the subjective constitution of our representative powers) an object of knowledge (*res cognoscibilis*), then our concepts will not be confronted

with objects, *but merely with our cognitive faculties* and the use which they can make of a given representation (in a theoretical or practical point of view). Thus the question whether anything is or is not a cognizable being is not a question concerning the possibility of things, but of *our* knowledge of them. (C.J., pp. 318–19, §91, at outset; my italics.)

### 3. The Categories of Evaluative Judgment and Practical Reasoning (Prudential and Moral)

The preceding section shows that on Kant's conception of the matter the forms of intuition and the categories of our understanding rest on the ultimately factual basis of the make-up of the human intellect. This circumstance also extends to embrace prudential reasoning and evaluative judgment.

Specifically, any application of the conceptions of means-to-ends and of purposes (*Ziele* and *Zwecke*) calls for a specification to the characteristically human frame of reference. Kant is emphatically explicit in insisting that purposiveness is a necessary feature of our human understanding and has a validity that stands strictly relativized to it.

Kant is emphatically explicit in insisting that purposiveness is a necessary feature of *our human understanding*:

> ... the concept of the purposiveness of nature in its products is *necessary for human judgment* in respect of nature, but has not to do with the determination of objects. It is, therefore, a subjective principle of reason for the judgment, which as regulative (not constitutive) is just as necessarily *valid for our human judgment* as if it were an objective principle. (C.J., pp. 252–253, § 76 *ad fin*; cf. p. 245, § 75; my italics.)

Accordingly, purposiveness is not a feature of the world "an sich," but is the product of a certain bias peculiar to the human mind in its conceptualization of "nature":

> Hence it is merely *a consequence of the particular constitution of our understanding* that it represents products of nature as possible, according to a different kind of causality from that of the natural laws of matter namely, that of purposes and final causes. (C.J., p. 256, § 77; my italics.)

Thus purposiveness is not something we learn from the study of nature, but something we bring to it in virtue of the intrinsic "constitution" of the human mind:

> Hence it is absolutely impossible for us to produce from nature itself grounds of explanation for purposive combinations, and *it is necessary by the constitution of the human cognitive faculties* to seek the supreme ground of these purposive combinations in an original understanding as the cause of the world. (C.J., p. 258, § 77 *ad fin*; cf. p. 267, § 80; my italics.)

The standing of purposiveness is consequently altogether "subjective" in being specifically correlative with the make-up of man's cognitive faculties. Moreover, the ultimate standing of the practical category of purposiveness is regulative:

> We can only assume this distinction [between causality in general and purposiveness in particular] as subjectively necessary by the constitution of our cognitive faculties and as valid for the reflective, not for the objectively determinant, judgment. But if we come to practice, then such a *regulative* principle . . ., which by the constitution of our cognitive faculties can only be thought as possible in a certain way, is at the same time *constitutive*, i.e., practically determinant. Nevertheless, as a principle for judging of the objective possibility of things, it is in no way theoretically determinant (i.e., it does not say that the only kind of possibility which belongs to the object is that which belongs to our thinking faculty), but is a mere *regulative* principle for the reflective judgment. (C.J., p. 309, § 88; cf. p. 265, § 79.)

And again:

> Hence the concept of an absolute necessary Being is no doubt an indispensable idea of reason, but yet it is a problematical concept unattainable by the human understanding. It is indeed *valid for the employment of our cognitive faculties in accordance with their peculiar constitution,* but not valid of the object. Nor is it valid for every knowing being, because I cannot presuppose in every such being thought and intuition as two distinct conditions of the exercise of its cognitive faculties, and consequently as conditions of the possibility and actuality of things. An understanding into which this distinction did not enter might say: All objects that I know *are*, i.e., exist; and the possibility of some which yet do not exist (i.e., the contingency of the contrasted necessity of those which do exist) might never come into the representation of such a being at all. But what makes it difficult for our understanding to treat its concepts here as reason does is merely that, for it, as human understanding, that is transcendent (i.e., impossible for the subjective conditions of its cognition) which reason makes into a principle appertaining to the object. Here the maxim always holds that all objects whose cognition surpasses the faculty of the understanding are thought by us according to the subjective conditions of the exercise of that faculty which necessarily attach to our (human) nature. If judgments laid down in this way (and there is no other alternative in regard to transcendent concepts) cannot be constitutive principles determining the object as it is, they will remain regulative principles adapted to the human point of view. . . . (C.J., pp. 250–251, § 76; cf. p. 308, § 88, and pp. 321–322, § 91.)

This purpose-oriented aspect of our understanding is accordingly contingent upon the specific constitution of our peculiarly human understanding:

> There emerges, therefore, a peculiarity of *our* (human) understanding in respect of the judgment in its reflection upon things of nature. But if this

be so, the idea of a possible understanding different from the human must be fundamental here. (Just so in the critique of pure reason we must have in our thoughts another possible [kind of] intuition if ours is to be regarded as a particular species for which objects are only valid as phenomena.) And so we are able to say: Certain natural products, from *the special constitution of our understanding, must be considered by us*, in regard to their possibility, *as if* produced designedly and as purposes. But we do not, therefore, demand that there should be actually given a particular cause which has the representation of a purpose as its determining ground, and we do not deny that an understanding, different from (i.e., higher than) the human, might find the ground of the possibility of such products of nature in the mechanism of nature, i.e., in a causal combination for which an understanding is not explicitly assumed as cause.

We have now to do with the relation of *our* understanding to the judgment, viz., we seek for a certain contingency in the constitution of our understanding, to which we may point as a peculiarity distinguishing it from other possible understandings. (C.J., pp. 253–254, § 77; the rest of this section is very relevant.)

Other sorts of intelligences whose concept of the "natural world" in which they operate is devoid of the aspect of purposiveness are certainly *in principle* conceivable, though how this possibility could be realized is something that is altogether inconceivable for us:

But for us men there is only permissible the limited formula: We cannot otherwise think and make comprehensible the purposiveness which must lie at the bottom of our cognition of the internal possibility of many natural things than by representing it and the world in general as a product of an intelligent cause [a God]. Now, if this proposition, based on an inevitable necessary maxim of our judgment, is completely satisfactory from every *human* point of view for both the speculative and practical use of our reason, I should like to know what we lose by not being able to prove it as also valid for higher beings, from objective grounds (which unfortunately are beyond our faculties). It is indeed quite certain that we cannot adequately cognize, much less explain, organized beings and their internal possibility according to mere mechanical principles of nature. . . . We must absolutely deny this insight to men. . . . So much only is sure, that if we are to judge according to what is permitted us to see by our own proper nature (the conditions and limitations of our reason), we can place at the basis of the possibility of these natural purposes nothing else than an intelligent Being. This alone is in conformity with the maxim of our reflective judgment and therefore with a ground which, though subjective, is inseparably attached to the human race. (C.J., pp. 247–248, § 75; cf. p. 251, § 76.)

The terminus of this finalistic purposiveness inherent in our understanding is God, conceived of as the ultimate causal source of purpose in nature. Seen from this angle, the status of necessity of a postulation of the deity is altogether coordinate with the general man-relativization of our apodeictic knowledge. In this concept of the deity, the practical category of purposiveness moves out to the ideal of reason that represents its ultimate vanishing point (*focus imaginarius*).

So much for the ultimately contingent and factual, man-relativized basis of our judgments of purposiveness in nature. It deserves remark that the prudential (i.e., non-moral) sector of our practical reasoning —which ultimately revolves about the recognition of the realization of happiness as the purpose specific to human life and activity[2]— rests on a foundation of exactly the same sort:

> It is clear, then, that it is *owing to the subjective constitution of our practical faculty* that the moral laws must be represented as commands and the actions conforming to them as duties, and that reason expresses this necessity, not by an "is" (happens), but by an "ought to be." . . . Though, therefore, an intelligible world in which everything would be actual merely because (as something good) it is possible, together with freedom as its formal condition, is for us a transcendent concept, not available as a constitutive principle to determine an object in its objective reality, yet *because of the constitution of our (in part sensuous) nature and faculty it is—so far as we can represent it in accordance with the constitution of our reason, for us and for all rational beings that have a connection with the world of sense—a universal regulative principle.* (C.J., pp. 251–252, § 76.)

The moral law in man and finality in nature are linked in indissoluble coordination:

> If we follow the latter order, it is a fundamental proposition, to which even the commonest human reason is compelled to give immediate assent, that if there is to be in general a *final purpose* furnished *a priori* by reason, this can be no other than *man* (every rational being of the world) *under moral laws*. . . . But the moral laws have this peculiar characteristic —that they prescribe to reason something as a purpose without any condition, and consequently exactly as the concept of a final purpose requires. The existence of a reason that can be for itself the supreme law in the purposive reference, in other words the existence of rational beings under moral laws, can therefore alone be thought as the final purpose of the being of a world. If, on the contrary, this be not so, there would be either no purpose at all in the cause of its being, or there would be purposes but no final purpose.
>
> The moral law, as the formal rational condition of the use of our freedom, obliges us by itself alone, without depending on any purpose as material condition, but it nevertheless determines for us, and indeed *a priori*, a final purpose toward which it obliges us to strive, and this purpose is the *highest good in the world* possible through freedom. (C.J., pp. 299–301, § 87; n.b. this whole section.)

As above, this necessity of subjection to moral law is a specifically human necessity inherent in the faculty structure of our minds:

[2] C.Pr.R., p. 20; Ak. 22.

> Pure reason, as a practical faculty, i.e., as the faculty of determining the free use of our causality by ideas (pure rational concepts), not only comprises in the moral law a regulative principle of our actions, but supplies us at the same time with a subjective constitutive principle in the concept of an object which [our] reason alone can think and which is to be actualized by our actions in the world according to that law. The idea of a final purpose in the employment of freedom according to moral laws has therefore subjective *practical* reality. We are *a priori* determined by reason to promote with all our powers the *summum bonum* [*das Weltbeste*], which consists in the combination of the greatest welfare of rational beings with the highest condition of the good in itself, i.e., in universal happiness conjoined with morality most accordant to law. (C.J., p. 304, § 88 at outset.)

Accordingly, Kant's theory of the necessity and universality of our synthetic *a priori* judgments is such that inevitably on both the theoretical and the practical side, it inheres in and derives from something that is ultimately a matter of brute fact, namely the "special constitution" of the human mind.

## 4. An Objection Overcome

It may seem at first sight that Kant's theory of morality forms an exception to the preceding account of his doctrine of universality and necessity. For in his discussion of the Categorical Imperative which provides the basis of his theory of morality, Kant emphatically says that this principle does not pertain to man alone, but holds good for *all rational creatures whatsoever:*

> Now this principle of morality, on account of the universality of its legislation which makes it the formal supreme determining ground of the will regardless of any subjective differences among men, is declared by reason to be a law for all rational beings in so far as they have a will, i.e., faculty of determining their causality through the conception of a rule, and consequently in so far as they are competent to determine their actions according to principles and thus to act according to practical *a priori* principles, which alone have the necessity which reason demands in a principle. It is thus not limited to human beings but extends to all finite beings having reason and will; indeed, it includes the Infinite Being as the supreme intelligence. (C.Pr.R., Bk I, ch. I, sec. 7 [Ak. p. 32].)

As this passage makes transparently clear, the universality and necessity of the moral law (viz., that *all . . . must*) is of a scope strictly universal for all rational creatures, and is not confined in applicability to man alone. And this seems at variance with the interpretation developed above.

It can be seen on closer scrutiny however, that this position does not in fact conflict with our general analysis of the man-relative basis of the universality and necessity of *a priori* synthetic truths. To see how a conflict is avoided we must revert to the distinction (of Sect. 1) between the *internal range of objects to which* a judgment pertains and the *external range of subjects for which* the judgment obtains. On my interpretation of Kant, this distinction becomes crucial at this juncture of the operative range of the CI (=Categorical Imperative). I take his position to be that the thesis "All rational minds must be subject to the CI" is akin to "All (plane) triangles must have interior angles summing to 180°" in that the former judgment (despite the fact of being about intelligences in general as its *internal* range) also has an *external* range that is confronted to the human sphere. The thesis is to be interpreted as "As we humans must of necessity judge the matter, all rational intelligences. . . ." The judgment holds *for* us men *of* all rational creatures, but need not of necessity hold *for* all rational creatures as such: it is not requisite that *all rational beings* hold it to be valid, but merely requisite that it be held valid by *us humans* for all rational beings. On this interpretation of the matter, the universality and necessity of moral thesis is no exception to the general rule of the anthropologism of the universality and necessity of synthetic *a priori* theses in Kant's system.

## 5. Russell's Critique of Kant and the Quest for Absolutism

In his classic work on *The Principles of Mathematics*, Bertrand Russell criticized Kant's theory of mathematical necessity in the following terms:

> Moreover, we only push one stage further back the region of "mere fact," for the constitution of our minds remains still a mere fact. The theory of necessity urged by Kant . . . appears radically vicious. Everything is in a sense a mere fact. A proposition is said to be proved when it is deduced from premisses; but the premisses, ultimately . . . have to be simply assumed. Thus any ultimate premiss is, in a certain sense, a mere fact.[3]

Russell thus objects that the Kantian theory of necessity gives a spurious account of necessitation, because in placing its ultimate source in the constitution of our minds it does not afford us any real necessity but hinges matters on an issue of "mere fact." Interestingly enough, this criticism coincides with Hegel's insistence (in §71 of the "*Lesser Logic*") that a philosophically adequate account of the certainty of something calls for showing that "this certainty, instead of proceeding from our particular mental constitution only, belongs to the very nature of mind."

[3] *The Principles of Mathematics* (London, 1903), p. 454.

This line of objection is highly questionable. There surely is no good reason why one should not be prepared, at any rate in the epistemological sphere, to acknowledge as adequate to our interests those modes of necessitation that are not absolute and altogether unconditioned but are relative and predicated upon the inherent structures of our cognitive mechanisms. Why should a necessity be dismissed as *spurious* that is not "purely logical" but has some degree of factual relativization? Our rationalist penchant for system and explanatory rationalization must not carry us away into asking for the whys of the whys (as Leibniz said of the Princess Sophie Charlotte of Hanover) to a point where we are utterly intolerant of any recourse to surd fact in the sphere of cognitive systematization. There must be some willingness, even in theoretical epistemology, to acquiesce in the ultimately factual basis of man's cognitive capacities and capabilities. Philosophers are surely not obliged to yearn for ultimacies to the point of feeling ashamed of an account of a specifically *human* mode of rationality that makes no pretense to cosmic validity from the perspective of any and every rational mind whatsoever.

This point has been driven home in 20th Century philosophy by the later work of Ludwig Wittgenstein. Wittgenstein emphatically stresses the ultimately factualistic and *irrationalizable* basis of our cognitive mechanisms in a "form of life": "What has to be accepted, the given, is—so one could say—*forms of life*."[4] (Recall Wittgenstein's dictum that if lions could talk we would not understand them.) For Wittgenstein, our conceptual necessities root in the ultimately factualistic basis of a language embedded in the operative setting of a "form of life"; for Kant, in the cognitive mechanisms inherent in the faculty structure of the human mind. (Thus for Wittgenstein this ultimate root is a *social*, while for Kant it is a *psycho-biological* reality: they are separated by the Hegelian revolution.) Kant and Wittgenstein thus stand together in opposing the necessitarian penchant of the "dogmatic" rationalists—to whose doctrine (specifically Spinoza's) the early Russell of *The Principles of Mathematics* was so powerfully attracted. This Wittgensteinian perspective on Kant's position regarding the ultimately factual roots of epistemic necessity seems to me useful in showing—not that it is right, but rather—that it represents a standpoint that is not only theoretically viable but at bottom neither eccentric nor even out of the style of contemporary philosophical fashion.

[4] *Philosophical Investigations* (Oxford, 1953), llxi, p. 226.

## 6. A Difficulty

Another difficulty remains. Someone might object:

> How are we to *learn* that the human understanding has certain factual (contingent) features except by empirical inquiry? And if our knowledge of facts must rest on experience what becomes of the whole system of claims to *a priori* truth?

To cope with this issue we must stress that "factual" does not entail "empirical," and that in this Kantian setting there will be other, non-empirical ways of bringing the factual basis of our necessary beliefs to light. The Kantian route here certainly proceeds not by *empirical* but by *analytical* means.

Kant's theory of truths whose foundation lies in the "special constitution" of man's mind is his functional equivalent of the theory of innate truths in Leibniz,[5] and represents what is perhaps the firmest link of the Critical Philosophy to the doctrines of its rationalist predecessor. In both cases, the "innateness" is a factual matter whose actuality is brought to light obliquely by *the logical analysis of our knowledge*, and not directly through recognition by some more direct, cognitive process (such as "experience"). What is crucial to Kant's position is that there must be some cognitive route to factual (and so synthetic) judgments that is *not* empirical, but somehow *a priori*. And this, of course, is quite natural on Kantian grounds. Is it not, after all, a task altogether central to the Critical Philosophy to provide by way of transcendental analysis the very means for realizing this objective? The critical task is not just to *maintain* that something is a necessary feature of the human mind, but to *exhibit* by analytical means just why and how it must be taken to be so.

The crucial consideration is that we are to move analytically from the nature of our *a priori* knowledge, as something "given," to the determination of the sort of factual foundation it requires. The reverse process *is simply not workable*. We cannot possibly move from the (somehow) *given* factual foundation for our *a priori* knowledge to an inferential determination of the specific items of *a priori* knowledge it validates (e.g., in geometry). And this is so for the simple reason that this "factual foundation" never appears in the role of an initial "given"; its status remains throughout that of an inferred entity. To adopt some very useful scholastic terminology, this factual foundation in human cognition provides the *ratio essendi* of our knowledge of

[5] For the Leibnizian theory and its ramifications see Chap. VII, "A New Look at the Problem of Innate Ideas" in Nicholas Rescher, *Essays In Philosophical Analysis* (Pittsburgh, 1969), pp. 255–270.

universal and necessary *a priori* synthetic truths; it certainly does not provide the *ratio cognoscendi* of this knowledge.

### 7. The Viability in Principle of the Kantian Theory of Necessity

Once one steps outside the realm of purely logical (analytical) necessity the question arises: Is there any other philosophical work left for the conception of necessity to do, does it retain any further domain of valid application? In general terms, the answer which contemporary modal theory affords is affirmative. The "party line" answer of contemporary philosophy runs as follows: Logical necessity is unconditional, absolute necessity, and beyond it there is also the conditioned, relative necessity (whose *formal* structure is altogether isomorphic with that of the former). This relativized mode of necessity is usually illustrated with reference to *physical* necessity, i.e., necessity relative to the (contingent) laws of nature as best we can determine them in the scientific investigation of the processes of nature. We accordingly arrive at a dual articulation of necessity: the "pure," unconditioned, logical version of the concept, and an "applied," conditioned, factually relativized version. (Both versions exhibit precisely the same abstract structure, and no one hesitates to recognize this law-relativized mode of necessitation as a legitimate version of the concept.)

Now there is no reason of principle why one could not contemplate a different sort of—fundamentally analogous—relativization, and consider a necessity relative not to the "laws of nature," but to the (contingent) *modus operandi* of the human mind as revealed in an analytical scrutiny of its cognitive products—and especially that of the conceptual scheme in whose coinage we transact our cognitive business (in science, in philosophy, and in everyday life). And this in principle perfectly feasible approach is precisely the line that the Kantian analysis takes. We might well for one reason or another feel disinclined to take this line, but there is nothing in the theoretical structure of the issues that would render it in principle infeasible. There is no reason of principle why the Kantian doctrine of necessary truths of the synthetic *a priori* type should yet be buried—nobody is in a position to certify to its demise.

V.

# Bertrand Russell and Modal Logic

## 1. Introduction

RUSSELL's repute as one of the founding fathers of modern symbolic logic is secure for all times and his claims to greatness as a logician are established to an extent beyond my meager capacity to alter for better or for worse. Accordingly, it is no real unkindness to Russell's memory to observe in the interests of historical justice that he too once more illustrates the rather trite precept that even scholars of deservedly great stature can exhibit a bias of intellect that produces unfortunate side effects. At any rate, the aim of this present discussion is to note the substantially negative import of Russell's work for the evolution of modal logic, whose rapid growth since the late 1940's is unquestionably one of the most exciting developments in contemporary logical research.

## 2. Philosophical Background

From the very first, Russell was on philosophical grounds reluctant, nay unwilling, to recognize the merely possible (i.e., the *contingently* possible) as a distinct category. His *Critical Examination of the Philosophy of Leibniz* (London, 1898) exemplifies this attitude. It was, Russell held, improper for Leibniz, given his own commitments, to espouse the category of mere possibility, and to maintain the contingency of factual truth: he should have held that all truths about the world are necessary. Thus Russell reproached Leibniz with not reaching more Spinozistic conclusions: had Leibniz traced out his own lines of thought more rigorously he would have arrived at the position of Spinoza.

Russell's criticism of Kant in *The Principles of Mathematics* (London, 1903) gives another revealing insight into his position. According to Russell, the Kantian analysis of the foundations of necessity is drastically insufficient. In tracing the source of necessity to the categories and forms of the human understanding, Kant—

so Russell holds—merely provides a contingently factual basis that cannot provide an appropriate foothold for necessity proper.

> [On the Kantian theory of necessity] we only push one stage farther back the region of "mere fact," for the constitution of our minds remains still a mere fact. The theory of necessity urged by Kant, and adopted . . . by Lotze, appears radically vicious. Everything is in a sense a mere fact. (*Op. cit.*, §430.)

The philosopher, Russell seems to imply, is engaged on a quest for the necessity of things that does not permit him to rest content, at any stage, with anything that is a matter of mere fact. Just this attitude lay behind Russell's rejection at this stage of the empiricist philosophy of mathematics of John Stuart Mill, which would not provide a suitable account of the necessity of mathematical truth.

The philosophical roots of the early Russell's discontent with merely factual truth are to be found in his prolonged flirtation with the philosophy of Spinoza, a marked feature of *Mysticism and Logic* and vividly at work in the splendid essay on "A Free Man's Worship." Drawn to Spinozistic necessitarianism on powerful ideological grounds, Russell shied away from all traces of Leibnizian possibilism.

Himself a determinist of more or less classical proportions, Russell was committed to a necessitarianism that left him disinclined on philosophical grounds to allocate a logically useful role to the modal distinctions between the possible, the actual, and the necessary. Like his hero, Spinoza, he was prepared to maintain that there will, in the final analysis, be a *collapse* of modality: that the actual itself is more or less necessary, so that the possible vanishes as a distinct category. This philosophical stance was, I believe, significantly operative in Russell's negative view *as a logician* regarding the utility and prospects of modal logic.

## 3. Mathematical Background

Russell's philosophical perspectives were, of course, substantially influenced by his preoccupation with mathematics. The early Russell pioneered the tendency, destined to become predominant in his later years, of approaching logico-philosophical problems from the mathematical point of view. Now mathematics has, of course, no place for modal distinctions: in mathematics it is altogether otiose to differentiate between the actual and the necessary, and there is no room at all for the contingently possible. Throughout the mathematical domain the drawing of modal distinctions is effectively beside the point.

Moreover, it seems particularly pointless to apply the concept of necessity to the theses of a mathematical system like Riemannian geometry; what is necessary—and also what is mathematically interesting—are the relationships of deductive consequence by which theorems follow from axioms. Accordingly, we find Russell maintaining in *The Principles of Mathematics* (London, 1903) that:

> Thus any ultimate premiss is, in a certain sense, a mere fact. . . . The only logical meaning of necessity seems to be derived from implication. A proposition is more or less necessary according as the class of propositions for which it is a premiss is greater or smaller. In this sense the propositions of logic have the greatest necessity, and those of geometry have a high degree of necessity. (§430.)

The *relative* necessity of mathematical propositions is to be defined in terms of implicative relationships, according as the body of the propositions that are needed as premisses for it is the less or that of propositions for which it can serve as premiss is the greater. And the necessity of deductive consequence is the basic mode of necessity as well as the only ultimately genuine form thereof. But now, once one follows Russell in defining pure mathematics as "the theory of propositions of the if-then form," one arrives at a view of pure mathematics that sees *all* its propositions as having this necessity of consequence. This strictly relativized necessity of deductive consequence is all one needs in the philosophy of mathematics, and so there is—for example—little point of speaking of the theorems of a mathematical system as necessary in ways other than as shorthand for "necessary relative to the axioms." Absolute and unrelativized necessity is not only otiose but obscurantist as well: the "necessity of consequence" is the ultimately basic form of necessity.

And at this point Russell's logicism intervenes decisively in the dialectic of thought. If the basis of our concern with logic is its role in the rational articulation of mathematics; nay, if there is at bottom a fundamental *identity* of logic with mathematics, then the handwriting is on the wall. For then if mathematics has no real need for modal distinctions and no room for contingent possibilities, then a modal *logic* becomes almost a contradiction in terms. This standpoint blocks any concern on the logicians part with mere possibilities and alternative possible worlds. Such concerns of traditional philosophy come to be seen as metaphysical sophistries upon which the logician must simply turn his back.

Thus, both from the point of departure of his philosophical determinism and of his mathematical logicism, Russell was powerfully predisposed against the maintenance of modal distinctions which could only secure their validation in a rationale that recognizes the prospect of contingent possibilities. These considerations provide the background for understanding Russell's relationship to those logicians—preeminently Hugh MacColl, C. I. Lewis, and Jan Lukasiewicz—who pressed for the recognition of modal distinctions during the period (roughly 1895–1925) when Russell was actively preoccupied with logic. Let us examine this phenomenon in some detail.

### 4. Interactions: (A) Russell and MacColl

In a series of articles published over a period of some 30 years beginning ca. 1880,[1] Hugh MacColl argued a number of points which any modern modal logician will recognize as foundational for his entire subject.

(1) that there is a crucial difference between propositions that obtain merely *de facto* and those that obtain of necessity; between those which *must* hold and those which *may or may not* hold (even if they actually do so). (The former type of truths MacColl characterized as *certain* the latter as *variable.*)

(2) that there is a crucial difference between a *material* implication and genuine implication. "For nearly thirty years," he complained in 1908, "I have been vainly trying to convince them [i.e., logicians] that this supposed invariable equivalence between a conditional (or implication) and a disjunction is an error."[2]

(3) that a satisfactory logic of modality must distinguish between actually existing individuals and merely possible ones; and that, accordingly, in constructing quantificational logic we should *not* simply and automatically presuppose that we are dealing with actually existing individuals.

Russell, of course, would have none of this. As far as he was concerned, all of MacColl's doctrines were the results of rather elementary errors. His distinction between certain and variable statements, for example, results from not distinguishing between propositions and propositional functions, and is simply a misguided and misleading way of dealing with the difference between them. There is no need to go beyond the twofold categorization of propositions proper as true and false.[3]

[1] MacColl published some 40 books and papers during the years from 1877 to 1910. For details see the bibliography by A. Church in vol. 1 (1936) of *The Journal of Symbolic Logic*; see pp. 132–233.

[2] " 'If' and 'Imply'," *Mind*, vol. 17 (1908), pp. 151–152 (see p. 152).

[3] " 'If' and 'Imply': A Reply to Mr. MacColl," *Mind*, vol. 17 (1908), pp. 300–301. Cf. *Introduction to Mathematical Philosophy* (London, 1919), p. 165.

In sum, Russell's philosophical positions and allegiances led him to dismiss all of MacColl's doctrines as so much old-fashioned fairytale nonsense.[4]

## 5. Interactions: (B) Russell and Meinong

One idea operative in MacColl—and even more prominently in the work of Alexius Meinong—came to arouse Russell's particular ire: the idea of unrealized or nonactual particulars. The conception that there are nonexistent individuals—i.e., particulars which don't exist in this, the actual world but could exist in some alternative dispensation—represents an idea of longstanding credentials in philosophy, figuring in the Presocratics, in medieval scholasticism, and in Leibniz, in addition to its prominent role in the philosophy of Brentano.[5]

According to Russell, MacColl's distinction between actual and merely possible individuals is the result of an incorrect theory of naming according to which any combination of letters which functions grammatically as a name must actually name something in virtue of this function. When in fact the name names nothing (e.g., "Pegasus")—and is thereby in Russell's opinion not properly speaking a name at all—MacColl, under the spell of linguistic usage, provides a referent for the name in the guise of a merely possible individual. For Russell, those theoreticians who, like MacColl and Meinong, accept an ontology of "merely possible objects" have fallen victim to the logically deceptive distortions inherent in our ordinary use of language.[6]

This conception of merely possible individuals is altogether anathema to Russell, who indeed flatly dismisses the very meaningfulness of ascribing existence to individuals. He writes:

[4] For Russell, the uncongeniality of MacColl's ideas was compounded by that of his somewhat idiosyncratic logical symbolism. The Russell Archives at McMaster University contain some 25 letters and postcards sent by MacColl to Russell over the years 1901–1909. In one of these MacColl complains that he and Russell have as much difficulty understanding one another as would an Englishman who knows little or no French and a Frenchman who knows little or no English. In a later handwritten annotation of a typed transcription of a letter of MacColl's dated May 28, 1905, Russell writes: "MacColl was a symbolic logician of some eminence. I have a very large number of letters from him, but I have not included them in this [typed] selection because they are in his difficult symbolism."

[5] For the history see the chapter on "The Conception of Nonexistent Individuals" in N. Rescher, *Essays in Philosophical Logic* (Pittsburgh, 1969).

[6] These considerations in the backwash of Russell's classic paper "On Denoting" (*Mind*, vol. 14 [1905], pp. 479–493)—that "paradigm of philosophy" as G. E. Moore called it, and as indeed it was for much of English philosophy during the inter-war era—are doubtless too familiar to need detailed documentation.

> For the present let us merely note the fact that, though it is correct to say "men exist," it is incorrect, or rather meaningless, to ascribe existence to a given particular $x$ who happens to be a man. Generally, "terms satisfying $\phi x$ exist" means "$\phi x$ is sometimes true"; but "$a$ exists" (where $a$ is a term satisfying $\phi x$) is a mere noise or shape—devoid of significance.[7]

Rather than run the risk of having to put up with nonexistent possible individuals, Russell is willing to dispense altogether with the whole process of attributing existence to things.

## 6. Interactions: (C) Russell and C. I. Lewis and J. Lukasiewicz

It is also illuminating to consider Russell's reactions to the work of two other pioneers of modal logic in its contemporary guise as a branch of symbolic logic. I think here primarily of C. I. Lewis, whose important *Survey of Symbolic Logic* appeared in 1918, and Jan Lukasiewicz, whose important historical and systematic inquiries came into increasing prominence after the early 1920's. It is a perhaps surprising, but, I think, interesting and not insignificant fact that one can search Russell's pages in vain for any recognition of the work of these men. (Russell's *Preface* to the second edition of the *Principia* [published in 1925] preserves total silence with respect to all these developments.) And this seems especially inexplicable in view of the fact that the issues that provided these writers with their entry-point into the realm of modal ideas were topics of very special interest to Russell. (In Lewis' case the motivating issue was the philosophy of Leibniz, in Lukasiewicz' it was that of determinism and problems of prediction and future contingency in the context of the philosophy of Aristotle.) Again, it is also startling that Russell also ignores totally the development of mathematical intuitionism, especially the writings of L. E. J. Brouwer, whose work provides a possible bridge to the modal realm from points of departure in the philosophy of mathematics.

A clear picture emerges: in pre-*Principia* days Russell sharply opposes philosophers like MacColl and Meinong who sought to promote concern with the logic of modalities; in post-*Principia* days, secure on his own logico-mathematical ground, Russell simply ignored writers like Lewis and Lukasiewicz and the institutionists, whose work could provide a basis for the introduction of modalities into the framework of symbolic logic. Where modal logic was concerned, Russell adopted Lord Nelson's precedent, and stolidly put his telescope to the blind eye.

[7] *Introduction to Mathematical Philosophy, op. cit.*

## 7. The Fascination with Truth-Functionality: Logical Atomism and Logical Positivism

The line of thought operative here is, of course, intimately linked to Russell's theory of logical constructions, and to the methodological precept of logical constructionism, which he articulates as follows:

> The supreme maxim in scientific philosophising is this: Whenever possible, logical constructions are to be substituted for inferred entities.[8]

Clearly, the dismissal of all inferred entities and processes points towards a demise of potentialities, powers, and causal efficacy that pulls the rug out from the main motivation for recognizing possibility and contingency. The logical construction of something real will, quite evidently, be a construction from elements that are themselves altogether actual (real). This part of Russell's philosophy provides yet another facet of his rejection of modality.

The stress upon logical reducibility was, of course, vastly congenial to the ethos of logical positivism. Taking increasingly definite form during the first decade after World War I, its programmatic menu offered a Hobson's choice between the *reduction* and the *abandonment* of philosophically problematic concepts. We are thus brought back to the influence of that "paradigm of philosophy" (according to G. E. Moore), the Theory of Descriptions, according to which some conception that is standardly operative in our ordinary scheme of thought about things is reductively annihilated as the mere product of linguistic illusion.

The generally reductivist penchant of the theory of logical constructions found its clearest expression in various aspects of Russell's reductivistic program in logic and the extraction of all logical operations from atomic elements by truth-functional modes of combination.

It is important to recognize that Russell was himself deeply caught up in the ideology of two-valued truth-functionality that was part of the heritage of Frege and received its canonical formulation in Ludwig Wittgenstein's *Tractatus Logico-Philosophicus*. The Russellian program of *The Philosophy of Logical Atomism*—as well as Wittgenstein's Tractarian theory correlated with it—envisaged the definitional reduction of all concepts of interest and utility in the precise sector of philosophy to truth-functional conjoinings or combinations

[8] Quoted in Rudolf Carnap, *Der logische Aufbau der Welt*: (Berlin, 1928), p. 1.

of basic propositions which (since meaningfully definite) will themselves be either true or false.[9] Being truth functional, these modes of combination have the feature that the truth-status of a compound can always be determined in terms of the respective status of its several constitutive components.[10]

Thus, significant weight came to be borne by not strictly logical but essentially methodological (perhaps even metaphysical) considerations as built into the philosophy of logical atomism. One simply lacks a rationally adequate grasp of theses that have not been analyzed into their components. Such analysis calls for indicating the component elements and composing structures through which the logical character of complexes is determined (*truth functionally* determined) in terms of the status of the component elements. If a thesis that is internally complex in its conceptual structure does not submit to truth-functional analysis, this is a mark of an internal imprecision whose toleration is a concession to obscurantism.

Now the critical fact which everyone alike recognized as a feature of modal concepts is that none of them—be they absolute (like possibility or necessity) or relative (like entailment or strict implication)—will be truth-functional. It was throughout recognized by all concerned as a vain enterprise to analyze modal concepts in two-valuedly truth functional terms. And for Russell and the bulk of the positivistically-inclined logical tradition that followed him down to the days of Goodman and Quine, this very fact provided the basis for a rejection of modality.

But from the first there were dissentients. The logical structure of the basic conceptual situation was created by an inconsistent triad:

(1) the insistence upon propositional two-valuedness;
(2) the insistence upon the truth-functionality of all proper propositional operators and connectives[11];
(3) the legitimacy of modal distinctions.

[9] But note that Wittgenstein gave hints from which Carnap developed his rationalization of modal logic.

[10] The strengths and limitations of this program were brought into clearest relief in A. Tarski's classic essay on the concept of truth in formalized languages ("Der Wahrheitsbegriff in den formalizierten Sprachen," Warsaw, 1930; original in Polish, German translation, 1933.) It is of interest in illustrating the pervasiveness of the truth-functional tenor of thought that Tarski in the early 1930's rejected the proposal (of Z. Zawirski and H. Reichenbach) to consider the probability calculus as a form of many-valued logic on the grounds that probabilities do not behave in a truth-functional manner. Cf. the discussion in N. Rescher *Many-valued Logic* (New York, 1969), pp. 184–188.

[11] There is no essential link between two-valuedness and truth functionality. Connectives in a many-valued logic can, of course, be truth-functional.

As indicated, Russell and his positivist congeners abandoned (3), but others took a different route. In his single-minded pursuit of a concept of relative necessity and a really viable analysis of if-then, C. I. Lewis gave up the truth functionality of (2), and developed the theory of strict implication. Jan Lukasiewicz in his pursuit of an Aristotelian theory of future contingency gave up (1), and developed many-valued logic. (Brouwer and his Intuitionist followers gave up the entire concept of mathematico-logical analysis upon which the concept of propriety operative in (2) is based.)

The ideological penchants and predilections of logical positivism involved: (i) a commitment to a sharp-edged criterion of truth that was unwilling to tolerate the pluralism of a theory of degrees of truth or to acknowledge—as apart from the altogether meaningless—any shades and gradations as between the true and the false, and (ii) a commitment to a criterion of meaning unwilling to recognize as meaningful conceptions not definitionally reducible to the clear conceptions of a canonical basis. Accordingly, while the appropriateness of efforts at a corresponding reduction of conceptions like absolute and relative modalities (or of counterfactual conditionals, to take another example) might be recognized, the failure of such a quest for two-valued reduction was taken as to be construed to spell not the inadequacy of the reductive program, but the illegitimacy of the putatively irreducible concepts. In this positivistic atmosphere, the Russellian distaste for modal concepts hardened into an attitude of virtually dogmatic rejection of modal logic.[12]

## 8. Effects

Orthodox two-valued and truth-functional logic—"classical" logic as it is now frequently called—in the form given it by Russell, his associates, and their followers, enjoyed enormous successes. The mainstream of development in the tradition of logicians like Hilbert, Gödel, Tarski, Church, Rosser, *et al.* developed logic into a powerful instrument for exploring the foundations of mathematics, and more than justified Russell and Whitehead's selection of that proud title of their monumental work.

Modal logic remained in the shadows for a long time. It did not really begin to come into its own until the development of modern

[12] The reluctance or inability of logical positivism to come to serious and effective grips with the logic of modality proved a serious stumbling block to the success of the movement. See the interesting essay by Hans Poser, "Das Scheitern des logischen Positivismus an modaltheoretischen Problemen," *Studium Generale*, vol. 24 (1971), pp. 1522–1535.

modal semantics, largely under the impetus of Rudolf Carnap[13] (erecting a structure of his own on foundations laid by Wittgenstein and Tarski). It was Carnap who first successfully elaborated the possible-world semantics which those who followed in his wake were to build up into the grandiose structure we know today. Until the late 1940's it remained to all intents and purposes the concern of a few eccentric philosophical guerillas concerned to snipe from the sidelines as the main column of modern mathematical logic marched by enroute from victory to victory in directions appointed for it by the orientation of Russell's work. Thus during the period from the early 1920's to the late 1940's the great bulk of logicians and logically-concerned philosophers—indeed virtually everyone outside the range of the personal influence of Lewis and Lukasiewicz[14] adhered to Russell's negative stance towards modal concepts. The great successes of the Russellian vision of logic in the mathematical sphere gave a massive impetus to his negative view of modality. The upshot was, I think it not unfair to say, that the development of modal logic was set back by a full generation.

There is no fundamental historical reason why modern symbolic modal logic could not have developed substantially sooner. The basic tools forged by MacColl and Lewis lay to hand by 1920, as did those hints of Wittgenstein's *Tractatus* (relating to probability) from which Carnap first systematized the possible-worlds interpretation of modal logic. There is no reason of historical principle why the logic of modality which surged up shortly after Carnap's *Meaning and Necessity* (1947) could not have begun soon after 1920. This development was certainly delayed by a full generation during the period between the two World Wars. This delay can be attributed in no small part to views and attitudes held by Russell and promulgated under the influence of his massive authority.

It would be just plain wrong to say that "the time was just not ripe" for the development of modal logic in the period between the two World Wars. The ideas were there, the pioneering work was being done, the relevant publications were part of the public domain. But this work simply did not have the reception it deserved—far too little attention was paid to it. And this was not due to any lack of intrinsic interest or importance or to any *logical* disqualifications, but principally because of *ideological* factors. Put bluntly, the de-

[13] *Introduction to Semantics* (Cambridge, Mass., 1942) and especially *Meaning and Necessity* (Chicago, 1947).

[14] Brouwer and the nonclassicists in the foundations of mathematics, of course, had no interest in modal logic as such, since modal concepts play no role in mathematics.

velopment of modal logic was retarded primarily because Russell and his positivist followers found modal conceptions *philosophically* uncongenial. And the influence of Russell was a crucially operative factor here. There is no question in my mind that if Russell had possessed a more urbane, tolerant, and receptive interest in logical work that did not resonate to his own immediate philosophical predilections, it would have done a great deal of good.

## 9. Assessment

Russell's work and the stimulus it exerted upon others was responsible for a massive forward step in the development of modern symbolic logic in its "classical" articulation, in a form eminently suited to mathematical developments and applications. The massive proportions of his contribution cannot be questioned. Nevertheless, insofar as the line of thought presented here is at all correct, it appears that baneful consequences ensued from Russell's work and its influence for the development of modal logic. But the question remains: Was this just an unfortunate historical accident or was it something for which Russell himself deserves a certain measure of responsibility?

This question is certainly not otiose or irrelevant. We know full well that missteps by later followers of a master must not inevitably be laid on his own doorstep. We cannot reproach the humane Dr. Guillotin—concerned only to minimize the agonies of criminals condemned to execution—with the excesses of the abuse of his favored implement during the Terror phase of the French Revolution. Nor can we reproach that dedicated and conscientious scholar Charles Darwin for the callous application of his ideas by some among the Social Darwinists. A master innovator can fall blameless victim to the rationally unbridled zeal and unrestrained excesses with which his followers exploit his work. He can certainly fall the unhappy hostage of an unforeseen and to him almost certainly unwelcome abuse of his ideas.

But is this defensive line available in Russell's case to blunt the charge of responsibility for impeding the development of modal logic? I think not. Because in this case the central factor is not question of the unforeseen and presumably undesired consequences of certain innovations, but of Russell's own views and positions. His own deliberately held negative views towards modal conceptions—opinions espoused on conscious and philosophically reasoned grounds—were themselves operative forces behind the impact of Russell's position in this sphere.

The distaste for modal logic in Anglo-American philosophy during the interim between the two World Wars was virtually initiated by Russell and largely propagated by his great influence. The development of modal logic was impeded neither by accidental factors nor because this branch of logic is itself lacking in substantive interest from a logical point of view, but because many logicians were led under Russell's influence to regard it as philosophically distasteful. It seems to me by no means unjust to place squarely at Russell's door a substantial part of the responsibility for the stunted development of modal logic during the two generations succeeding the pioneering days of Hugh MacColl. Russell's philosophically inspired attitudes propagated a negative view of modal logic and helped to produce that disinclination to take modality seriously which can still be seen at work among our own contemporaries of the older generation (e.g., W. V. Quine and N. Goodman). The very success of Russell's work in the more mathematically oriented sectors of logic gave authority and impact to his antagonistic stance towards the logic of modality. For the development of *this* area of logic, at any rate, Russell's work represented a distinctly baneful influence.

Please do not misunderstand my intentions. It is not my aim to accuse Russell of any sort of wickedness in regard to modal logic. After all, every philosopher is entitled to his full share of human failing, myopia, and even prejudice. My concern is not so much with moral as with causal responsibility. It is my prime aim to persuade you of the *causal* fact that Russell's disinclination towards modal conceptions substantially retarded the development of modal logic. As regards issue of praise or blame I leave it to the reader to draw his own conclusions.

Certainly nothing could be more wise and urbane than the pious sentiments of the concluding paragraph of Russell's review in *Mind* of MacColl's *Symbolic Logic and Its Applications:*

> The present work . . . serves in any case to prevent the subject from getting into a groove. And since one never knows what will be the line of advance, it is always most rash to condemn what is not quite in the fashion of the moment.[15]

Anyone concerned for the health and welfare of modal logic as an intellectual discipline cannot but wish that Russell himself—and especially that majority among his followers who were perhaps even more royalist than their king—had seen fit to heed this eminently sound advice.

[15] *Mind*, vol. 15 (1906), pp. 255–260 (see p. 260).

# Part 2:
# SYSTEMATIC STUDIES

# VI.

# On Alternatives in Epistemic Logic

## 1. Introduction

THIS essay will study the systematization of certain epistemic principles of inference and survey various systems of epistemic logic embodying these principles. This survey is not intended to be exhaustive, nor is it claimed that the principles and systems presented are the somehow "correct" ones. Quite to the contrary, we seek to emphasize that, because of the fundamental diversity of plausible epistemic principles, the task of epistemic logic must be conceived as that of presenting and exploring various *alternative* systems.

It is well to observe at the outset that epistemic logic cannot concern itself with actual *occurrent* knowledge, nor with *dispositional* knowledge: these biographical and psychological approaches to knowledge simply lack a "logic." For logic must deal with inferences, and even so simple and straightforward an inference as that from someone's knowledge of $p$-and-$q$ to his knowledge of $p$ would not be feasible here—though he knows $p$-and-$q$ "occurrently," $p$ as such might never have occurred to him separately. (And an analogous situation obtains in the dispositional case.) If we took this strictly empirical approach we would be thrown back upon E. J. Lemmon's pessimistic view that the sole acceptable principle in epistemic logic is the inference from someone's knowledge of something to its truth. The price that epistemic logic must pay for an insistence upon overt (occurrent or dispositional) knowledge is that of vacuity.

Epistemic logic must deal with *virtual* or *implicit* knowledge, that knowledge which a knower can *in principle* come to have, what a knower can *in principle* determine to be the case given with what he—explicitly or implicitly—knows. But the question of what, *in principle*, a knower can come to know is a question that epistemic logic itself cannot answer in any univocal and unequivocal way. For the answer to this question will depend on what we shall recognize to be *logically* valid principles of inference in this sphere. Since this is an inherently pluralistic issue, it would appear that epistemic logic cannot but shoulder the task of presenting a diversity of systems.

## 2. Principles of Epistemic Inference

This section will consider a number of epistemic principles, principles in a sense selected arbitrarily. That is, they are arbitrary with respect to the full range of principles that one *might* have considered at this point. Yet, these principles are not arbitrary in themselves, because they articulate to some extent our somewhat loose, informal intuitions concerning *implicit* knowledge. And precisely because our intuitions in this matter lack a monolithic fixity and *determinacy* with respect to favoring one principle over against another in a decisive way, we consider it an inescapable position that there is not one all-embracing mode of implicit knowledge, but rather that there are various modes of implicit knowledge, each mode giving rise to a distinct system of epistemic logic (and *vice versa*).

The propositions with which epistemic logic deals take the form "$x$ (the knower) knows that $A$" and the various components thereof. Accordingly, the prime task of any system of epistemic logic is to specify what formal relationships hold among propositions of this sort. In what follows we shall write "$KxA$" for "$x$ knows that $A$" and we shall use '$\sim$,' '&' and '$\rightarrow$' for negation, conjunction, and implication, respectively. (We need not assume that the implication at issue is *material* implication but could assume it represents a stronger relationship, such as entailment.)

The one absolutely undisputed principle of epistemic logic—which characterizes even overt knowledge—is that one cannot know what is false:

(1) $\dfrac{KxA}{A}$ "What is known must be the case"

It can plausibly be held that another principle must also characterize the knowledge relationship:

(2) $\dfrac{Kx(A\rightarrow B),\ KxA}{KxB}$ "What is known to follow from the known is itself known"

This would be argued on the ground that if someone knows that $A$ and did not know that $B$, then it would be implausible to maintain—and we would be disinclined to concede—that he *knows* that $A$ implies $B$. (Yet, such anthropological considerations are not conclusive.) We would, at any rate, require that (2) holds for *implicit* knowledge, that is, require that *modus ponens* holds for what one knows.

Given a system of logic **S** that is to provide the non-epistemic basis for the development of an epistemic logic, we may go beyond the minimum requirement of (1) and (2) by strengthening the concept of implicit knowledge to that of knowing all the logical consequences (with respect to the system **S**) of what one knows. We would then have the principle:

(3) $\dfrac{KxA}{KxB}$, provided $\vdash_s A \rightarrow B$ "What follows *logically* from the known is known"

We note that if **S** contains the theorem $B \rightarrow A$, for any $B$, whenever $A$ is a theorem, then in the presence of (2) and the assumption that the knower $x$ knows anything at all, (3) implies that one knows all the logical truths of **S**, that is,

(4) $KxA$, provided $\vdash_s A$ "What is *logically* demonstrable is always known"

That (4) and (2) yield (3) in return is obvious.

To arrive at such a conception of implicit knowledge in the most natural way, we could well start with a *strict* concept of knowledge, $K^*xA$—where $K^*$ represents, say, actually occurrent knowledge—and then define

$$KxA \text{ iff for some } B,\ K^*xB \text{ and } \vdash_s B \rightarrow A$$

i.e., something is *implicitly* known if it is an **S**-consequence of something that is known in the strict sense. Such a definition satisfies (3) provided that the transitivity-principle inference

$$\vdash_s A \rightarrow B,\ \vdash_s B \rightarrow C, \text{ therefore } \vdash_s A \rightarrow C$$

holds for **S**. Moreover, (4) also obtains if the inference

$$\vdash_s A, \text{ therefore } \vdash_s B \rightarrow A$$

holds for **S**, as was noted above.

A further strengthening of the concept of implicit knowledge can be achieved by strengthening (3) to

(3′) $\dfrac{KxA}{KxB}$ provided $\vdash A \rightarrow B$ "What follows from the known is known"

i.e., by dropping the requirement that *$A \rightarrow B$ be provable* in **S** and requiring instead that *$A \rightarrow B$ be provable in the system of epistemic logic* for which (3′) is itself a rule. Again, such a strengthening would be tantamount to adding the principle

(4′) $KxA$, provided $\vdash A$ "What is demonstrable is known"

There is yet another dimension of variation of implicitness which has not to do with logical consequences of what is known but with knowledge regarding what is known. Let us consider inferences of the form:

(F) $\frac{A}{KxA}$ provided $A$ satisfies the condition $C$.

One condition on $A$ can be that $A$ itself is of the form $Kx(\ldots)$, so that we would have the principle

(5) $\frac{KxA}{Kx(KxA)}$ "What is known is known to be known"

Such inferences reflect the intuitive idea that "what someone knows, he knows that he knows." Here one might picture someone searching the *complete* (infinite) inventory of the propositions stating what he knows. If such a search can be carried out, then $KxA$ must also be on the inventory whenever $A$ is on it.

Another (far less plausible) condition upon $A$ in ($F$) might be that $A$ take the form $\sim Kx(\ldots)$, leading to

(6) $\frac{\sim KxA}{Kx(\sim KxA)}$ "What is unknown is known to be unknown"

Again, one might here picture someone inspecting the inventory of what he knows. The difference between this case and the previous one is that in this case the inspector must *know* that the inventory is complete, so that if he does not find $A$ in the inventory, then he will know that he does not know that $A$—in fact, and in that case it must be that $\sim KxA$ is registered in it.

Principles (2)–(6) and their variants are all closely related principles. We have already observed the interrelatedness of (2)–(4). Note now the close relationship between (3′) and (5). If we consider the intuitive principle:

(R) One knows what follows from his knowledge

we see that (R) is thoroughly ambiguous in that it can be rendered in the following ways:

(R1) $\frac{KxA}{KxB}$ provided $\vdash A \rightarrow B$

(R2) $\frac{KxA}{KxB}$ provided $\vdash KxA \rightarrow B$

(R1), which is (3′) says that one knows what follows from *what* he knows. (R2) however, says that one knows what follows from *his knowing* what he knows. These principles are clearly different. However, it turns out that (R1) is interdeducible with the conjunction of (R1) and (5) (i.e.: (R2)$\Leftrightarrow$[(R1) & (5)]). That (R2) implies (5) is immediate, since we must inevitably have $\vdash KxA \rightarrow KxA$. That (R2) implies (R1) can be seen as follows:

1. $KxA$ (where $\vdash A \rightarrow B$) hypothesis

(But, as explained earlier, if $\vdash (A \rightarrow B)$ then $\vdash KxA \rightarrow (A \rightarrow B)$.)

2. $Kx(A \rightarrow B)$ 1, (R2)
3. $KxB$ 1, 2, (2)

To see that (R1) and (5) yield (R2) we note that

1. $KxA$ (where $\vdash KxA \rightarrow B$) hypothesis
2. $Kx(KxA)$ 1, (5)
3. $KxB$ 1, 2, (R1)

Thus, in the presence of (2) and an Irrelevance Principle to the effect that if $\vdash D$ then $\vdash C \rightarrow D$, (R2) is equivalent with (R1) and (5).

By an argument analogous to the preceding, it is possible to show that the (far less plausible) principle:

One knows what follows from his ignorance

if construed as

(R3) $\dfrac{\sim KxA}{KxB}$ provided $\vdash \sim KxA \rightarrow B$

is in fact equivalent with the conjunction of (R1) and (6) (i.e., (R3)$\Leftrightarrow$ [(R1) & (6)], given (2) and the Irrelevance Principle.

## 3. Epistemic Systems in Axiomatic Perspective

We shall now consider some systems of epistemic logic that arise by the embodiment of the principles considered above. In their construction we shall begin with a standard system of propositional logic —say (for ease and convenience) the two-valued propositional calculus **PC**. The basic starting-point system of alethic propositional logic will be supplemented by adding to it some of the principles (1)–(6). The systems thus obtained—and even their mode of formulation—will not be unfamiliar. Because the epistemic principles we have selected are the exact counterparts of standard alethic *modal* principles, the systems we consider will be the exact epistemic counterparts of standard modal calculi.

Each of the systems below will have as a primitive basis the propositional variables and the connectives '$\sim$,' '&,' and '*Ka*.' The other connectives are defined in the usual way. We shall use upper case letters as variables over formulas and shall use '$\Leftrightarrow$' to indicate the inference authorized by the inference rules.

We shall conveniently refer to the following family of three rules as **PC**:

(MP) $\vdash A, \vdash A \supset B \Rightarrow \vdash B$
(Subst.) Uniform substitution of formulas for variables
(PC) $A$ is a two-valued tautology $\Rightarrow \vdash A$

(I) The epistemic system **0** is **PC** plus

(K1) $\vdash Kap \supset p$ [cf. rule (1) of the preceding section]
(K2) $\vdash Ka(p \supset q) \supset (Kap \supset Kaq)$ [cf. rule (2)]

(II) The system **S0.5** is **0** plus
(KPC) $\vdash_{PC} A \Rightarrow \vdash KaA$ [cf. rule (4)]

(III) The system **T** is **S0.5** plus
(K′) $\vdash A \supset B \Rightarrow \vdash KaA \supset KaB$ [cf. rule (3′)]

we note that **T** is thus equivalent to **0** plus
(K) $\vdash A \Rightarrow \vdash KaA$ [cf. rule (4′)]

(IV) The system **S4** is **T** plus
(K4) $\vdash Kap \supset Ka(Kap)$ [cf. rule (5)]

(V) The system **S5** is **T** plus
(K5) $\vdash \sim Kap \supset Ka(\sim Kap)$ [cf. rule (6)]

As the designation of these systems suggests, each is in fact the epistemic translation of an already familiar system of alethic modal logic. Thus by suitably sequential introduction of the epistemic rules discussed in Section II, it is possible to build up a series of systems of *epistemic* modal logic in essential parallelism with the standard alethic case.

On our view there is no one single and uniquely "correct" system of epistemic logic, and the quest for it seems misguided. (The present approach to epistemic logic thus differs from the position of J. Hintikka's *Knowledge and Belief* [Ithaca, 1962], where a single system of epistemic logic—essentially **S4**—is put forward and defended). On the present position, "to know" has a variety of distinct albeit related senses (e.g., "occurrent" knowledge *vs.* "implicit" knowledge in the several very different modes of implicitness), different ones of which are governed by different constellations of rules. The study of epistemic logic, so viewed, is the inherently pluralistic study of diverse epistemic systems.

## 4. Epistemic Systems in Semantical Perspective: Epistemic "Possible Worlds"

The notion of possible worlds has been a highly fruitful concept in modal logic and can play an important role in epistemic logic as well. Apart from its role in proofs of completeness, the possible world semantics provide us with an advantageous vantage point from which to view epistemic logic: though it may appear as an inviting prospect that the diversity we have encountered on the axiomatic side of epistemic logic can be removed by considerations on the semantical side, we propose to show, in the remainder of the paper, that this prospect is unlikely to be realized—the axiomatic diversity can be mirrored semantically.

We can think of possible worlds as (possible) situations in which certain states of affairs obtain, so that certain propositions—those asserting these states of affairs—are true. We can, therefore, in fact identify a possible world with the set of propositions true in it. Propositions that characterize possible epistemic worlds are propositions not only about what is "objectively" the case but also about what is known or not known in those worlds (and about relationships between such propositions). What relationships between propositions hold in those worlds will in turn depend on the mode of knowledge that is at issue. If we consider a very weak mode of implicit knowledge, then for *all* epistemic worlds $\lambda$, if $Ka(p \,\&\, q)\epsilon\lambda$ and $Ka(p \,\&\, q \supset p)\epsilon\lambda$ then $Kap\epsilon\lambda$, even though for some world $\kappa$ it might be that $Ka(p \,\&\, q)\epsilon\kappa$ and $Kap\notin\kappa$, if $Ka(p \,\&\, q \supset p)\notin\kappa$. But this could not be the case for a stronger mode of implicit knowledge. What counts as a possible epistemic world, then, depends on what epistemic mode is at issue. (The problem can, of course, be seen in the reverse manner: what epistemic mode is at issue will depend on what possible worlds are like.)

## 5. The Gamut of Possible Worlds

We now turn to the problem of characterizing possible epistemic worlds corresponding to the various modes of implicit knowledge that we discussed. First, some notation must be introduced.

Whenever convenient, we shall use in our metalanguage the universal and existential quantifiers '$\forall$' and '$\exists$,' over both formulas and sets, and '$\Rightarrow$' and '$\Leftrightarrow$' for "if, then" and "if and only if," respectively. We shall use the letters '$A$,' '$B$,' and '$C$' as variables over formulas, and '$\gamma$,' '$\kappa$,' and '$\lambda$' as variables over sets. Also, we shall generally simplify "$KaA$" to "$KA$." The usual notions and notation of set theory will be employed: in particular,

$$\bigcup \lambda = \{A:(\exists\kappa)\,(\kappa\epsilon\lambda \text{ and } A\epsilon\kappa)\} = \bigcup_{\kappa\epsilon\lambda} \kappa$$

$$\bigcap \lambda = \{A:(\forall\kappa)\,(\kappa\epsilon\lambda \Rightarrow A\epsilon\kappa)\} = \bigcap_{\kappa\epsilon\lambda} \kappa$$

In the discussion that follows the following three definitions will be useful:

$$\mathbf{S}^* = \{A:A \text{ is provable in } \mathbf{S}\}$$
$$[\lambda] = \{A:KA\epsilon\lambda\}$$
$$<\lambda> = \{A:\sim K\sim A\epsilon\lambda\}$$

So much for notational preliminaries. We now proceed to the definition of the key concepts around which our deliberations will revolve.

The first definition introduces the idea of a possible world, construed (informally speaking) as the set of all propositions true in a possible world rather than as a single conjunctive proposition uniting all these truths:

(D1) $\lambda$ is a **PC**-*world*$\Leftrightarrow$(1) $\lambda$ is *maximal*: $A\epsilon\lambda$ or $\sim A\epsilon\lambda$
(2) $\lambda$ is *consistent*: $A\notin\lambda$ or $\sim A\notin\lambda$
(3) $\lambda$ is a *filter*: $A \,\&\, B\epsilon\lambda \Leftrightarrow (A\epsilon\lambda$ and $B\epsilon\lambda)$

The characterization of **PC**-worlds is one that we would require of *any possible* state of affairs, epistemic or otherwise. We note that in the presence of the (D1.1) and (D1.2), (D1.3) can be replaced by (i.e., is equivalent with the conjunction of)

(4) $\lambda$ is *mp-closed* (closed under modus ponens): $A\epsilon\lambda$ and $A\supset B\epsilon\lambda \Rightarrow B\epsilon\lambda$
(5) $\lambda$ is **PC**-*containing*: $\mathbf{PC}^*\subseteq\lambda$

Our next definition adds some further specifications to the preceding:

(D2) $\lambda$ is an **O**-*world*$\Leftrightarrow$1) $\lambda$ is a **PC**-world
(2) $[\lambda]\subseteq\lambda$
(3) $[\lambda]$ is mp-closed

Condition (D.2.2) requires of **O**-worlds $\lambda$ that if $KA\epsilon\lambda$ then $A\epsilon\lambda$, so that $KA \supset A \epsilon \lambda$, and condition (D2.3) requires that $K(A \supset B) \supset (KA \supset KB) \epsilon \lambda$. Our characterization of **O**-worlds thus amounts to:

$\lambda$ is an **O**-world $\Leftrightarrow$ (i) $\lambda$ is maximal, consistent, and mp-closed
(ii) $\mathbf{O}^* \subseteq \lambda$

(D3) $\lambda$ is an **S0.5**-*world* $\Leftrightarrow$ (1) $\lambda$ is an **O**-world
(2) $\mathbf{PC}^* \subseteq [\lambda]$

What distinguishes **O**-worlds from **S0.5** worlds is that in the latter all two-valued tautologies are *known* (i.e., present in *K*-modalized form), whereas in the former this is not required. Again, it follows immediately that the following characterization is equivalent to (D3):

$\lambda$ is an **S0.5** world $\Leftrightarrow$ (i) $\lambda$ is maximal, consistent, and mp-closed
(ii) $\mathbf{S0.5}^* \subseteq \lambda$

To characterize T-worlds, and worlds containing T-worlds, we must take into account the fact that T has the *rule of epistemization*: $\vdash A \Rightarrow \vdash KA$. Accordingly, we shall reflect this rule explicitly in our definition of T-worlds. (Later, we shall give an alternative characterization of T-worlds without explicit resort to this rule.)

(D4) $W$ is a **T**-*world structure* $\Leftrightarrow$
(1) $\lambda \epsilon W \Rightarrow \lambda$ is an **O**-world
(2) $A \epsilon \bigcap W \Rightarrow KA \epsilon \bigcap W$

(D5) $\lambda$ is a T-world $\Leftrightarrow$ for some T-world structure $W$, $\lambda \epsilon W$.

Thus, **T**-worlds are **O**-worlds which contain the epistemization of propositions which are true in all the appropriately related T-worlds (as specified in D4.2).

(D6) $W$ is an **S4**-*world structure* $\Leftrightarrow$
(1) $\lambda \epsilon W \Rightarrow \lambda$ is an O-world and $[\lambda] \epsilon [[\lambda]]$
(2) $A \epsilon \bigcap W \Rightarrow KA \epsilon \bigcap W$

(D7) $\lambda$ is an **S4**-*world* $\Leftrightarrow$ for some **S4**-world structure $W$, $\lambda \epsilon W$

Condition (D6.1) requires of every **S4**-world $\lambda$ that $KA \supset KKA \epsilon \lambda$.

(D8) $W$ is an **S5**-*world structure* $\Leftrightarrow$
(1) $\lambda \epsilon W \Rightarrow \lambda$ is an **O**-world and $[\langle \lambda \rangle] \subseteq [\lambda]$
(2) $A \epsilon \bigcap W \Rightarrow KA \epsilon \bigcap W$

(D9) $\lambda$ is an **S5**-*world* $\Leftrightarrow$ for some **S5**-world structure $W$, $\lambda \epsilon W$.

Here (D8.1) guarantees that the **S5**-thesis $\sim KA \supset K \sim KA$ belongs to every **S5**-world.

The characterization of these different kinds of worlds for various systems **S** (**S**-worlds) that we have given are by no means arbitrary and without *arrière pensée*. They have throughout been motivated by the intuitive general principle that:

$$\lambda \text{ is an } \mathbf{S}\text{-world} \Leftrightarrow \text{(i) } \lambda \text{ is maximal, consistent, and mp-closed}$$
$$\text{(ii) } \mathbf{S}^* \subseteq \lambda$$

That this principle holds for **PC**-, **O**-, and **S0.5**-worlds has already been noted. It will be shown in the Appendix that this principle also holds for **T**-, **S4**-, and **S5**-worlds as well. It is also demonstrated there that:

$$A \epsilon \mathbf{S}^* \Leftrightarrow A \epsilon \bigcap \{\lambda : \lambda \text{ is maximal, consistent, mp-closed, and } \mathbf{S}^* \subseteq \lambda\}$$

Accordingly, we can obtain the completeness result:

$$A \epsilon \mathbf{S}^* \Leftrightarrow \text{for every } \mathbf{S}\text{-world } \lambda,\ A \epsilon \lambda$$

It is possible to give alternative characterizations of the epistemic worlds of the preceding discussion. For present purposes we need the conception introduced by the recursive definition:

$$[\lambda]^0 = \lambda;\ [\lambda]^{i+1} = [[\lambda]^i]$$

We consider now the following conditions:

(1) $\lambda$ is maximal, consistent, and mp-closed
(2) $\mathbf{PC}^* \subseteq [\lambda]^i$
(3) $KA \supset A \epsilon [\lambda]^i$
(4) $K(A \supset B) \supset (KA \supset KB) \epsilon [\lambda]^i$
(5) $KA \supset KKA \epsilon [\lambda]^i$
(6) $\sim KA \supset K \sim KA \epsilon [\lambda]^i$

We leave it to the reader to verify that the following sequence of system-characterizations present alternative formulations that are wholly equivalent to the ones previously given:

$\lambda$ is a **PC**-world $\Leftrightarrow \lambda$ satisfies 1, 2 ($i=0$)
$\lambda$ is an **O**-world $\Leftrightarrow \lambda$ satisfies 1, 2–4 ($i=0$)
$\lambda$ is an **S0.5**-world $\Leftrightarrow \lambda$ satisfies 1, 2 ($i=1$), 3–4 ($i=0$)
$\lambda$ is a **T**-world $\Leftrightarrow \lambda$ satisfies 1, 2–4 (for all $i \geqslant 0$)
$\lambda$ is an **S4**-world $\Leftrightarrow \lambda$ satisfies 1, 2–5 (for all $i \geqslant 0$)
$\lambda$ is an **S5**-world $\Leftrightarrow \lambda$ satisfies 1, 2–6 (for all $i \geqslant 0$)

This way of viewing the matter makes it perfectly clear that the particular sorts of worlds on which we have focussed (and the specific systems associated with them) are but a handful of special cases chosen from among a gamut of possible alternatives.

## 6. The Introduction of Accessibility

Possible world semantics came to a significant advance with the introduction of an *accessibility*, or *alternativeness relation* among possible worlds. The intuitive idea behind such a relation is that certain worlds are *alternatives to*, or *accessible to*, or are *possible relative to* other worlds, while others are not. Let us denote this alternativeness relation by 'R' and read "$\lambda R\kappa$" as "$\kappa$ is an alternative (world) to $\lambda$." Customarily, a possible world semantics is approached with the relation $R$ as a primitive notion subject to certain conditions and, in particular, to the condition:

$A$ is $K$-modalized (in our case *known*) in a world $\lambda\Leftrightarrow$
$A$ is *true* in every alternative world to $\lambda$

that is,

$$\lambda\in W\Rightarrow[KA\in\lambda\Leftrightarrow(\forall\kappa)\ (\kappa\in W \text{ and } \lambda R\kappa\Leftrightarrow A\in\kappa)]$$

where $W$ is some (unspecified) set of (unspecified) worlds. On this approach, what worlds are like depends on what the relation $R$ is like.

In the spirit of this approach, we can define a relation $R$ over the domain of S-worlds as follows:

$$\lambda R\kappa\Leftrightarrow[\lambda]\subseteq\kappa$$

i.e., an S-world $\kappa$ is an *alternative to* an S-world $\lambda$ (in an epistemic sense of the term) just in case everything that is known in $\lambda$ is in fact true in $\kappa$.

These two notions of alternatives are not equivalent, since if $R$ is primitive, one cannot guarantee

$$\kappa,\ \lambda\in W\Rightarrow([\lambda]\subseteq\kappa\Rightarrow\lambda R\kappa)$$

and, if S-worlds are primitive, one cannot guarantee that for any S-world $\lambda$

$$(\forall\kappa)\ (\kappa \text{ is an S-world and } \lambda R\kappa\Rightarrow A\in\kappa)\Rightarrow KA\in\lambda.$$

That this is so can be seen directly from the fact that for a system **S** which does not have the rule of epistemization we have $A\in\mathbf{S}^*$ and $KA\notin\mathbf{S}^*$, for some formulas $A$, so that there are S-worlds not containing $KA$, even though *every* S-world (and hence every *alternative*) does contain $A$. In the appendix, however, we show that for systems **S** for which the rule of epistemization holds we have the result that for any S-world $\lambda$:

$$KA\in\lambda\Leftrightarrow(\forall\kappa)\ (\kappa \text{ is an S-world and } \lambda R\kappa\Rightarrow A\in\kappa)$$

If, on one's intuitions, one adopted the view that if $A$ is not known, then $A$ is false in some epistemic alternative, then, indeed, for certain systems **S** not containing the rule of epistemization, one would have to recognize certain 'impossible' **S**-worlds, e.g., worlds in which $A \supset A$ were false. This view would afford, then, a criterion by which to judge acceptable systems of epistemic logic. For example, on this criterion the systems **0** and **S0.5** would be unacceptable since for some **O**-worlds $\lambda$ we have that $K(A \supset A) \notin \lambda$, so that for some alternatives $\kappa$ to $\lambda$, $A \supset A \notin \kappa$; and since for some **S0.5**-worlds $\lambda$ we have that $K(KA \supset A) \notin \lambda$, so that for some alternatives $\kappa$ to $\lambda$, $KA \supset A \notin \kappa$. It does not seem, however, that this criterion actually goes so far as to *force* one to accept only systems at least as strong as **T**. All the same, the key role of the rule of epistemization with regard to the accessibility principle does set this rule in a critical light.

The deliberations of this and the preceding section indicate a close parallelism, in the construction of systems of epistemic logic, between the axiomatic approach through epistemic principles and the semantic approach through epistemic worlds. This parallelism destroys all hope that in the present epistemic case the pluralism met on one side of the axiomatic/semantic divide can be overcome by a more rigid monism upon the other.

## 7. Conclusion

The survey of systems carried out here substantiates the pluralistic position stated at the outset. There is clearly a great variety of epistemic systems, depending on just exactly how we choose to articulate the basic 'logical' rules from which the systems grow.

Our approach to epistemic logic thus emphasizes pluralism. All the same, it does seem possible to delimit the range of acceptable systems at least to some extent: certain systems are clearly too weak, and others are too strong. A system containing $A \supset KA$ as a thesis is too strong since it requires knowers to be omniscient. Also, the system **S5** appears to be too strong in that it requires knowers to have knowledge concerning their own ignorance, and this seems to require too much. (To be sure, if a knower were to be able to inventory *all knowable statements*, he could then, of course, go through the list to find out—and thus know—what he does and does not know.) At the other end of the spectrum of epistemic systems, we would certainly require that systems be at least strong enough to contain the thesis $KA \supset A$, and to allow for the implicit knowledge of truths that are not explicitly known.

Finally, our intuitions regarding rules basic to the logic of knowledge are certainly not all that precise. Minor variants of these rules are readily available which differ only subtly and unobtrusively in point of their "intuitive acceptability" but whose substitution for one another may lead to major *systematic* differences. We encounter here a conceptual analogue to a physically unstable situation where small perturbations lead to large resulting effects. In our case, seemingly small differences among the rules can correspondingly amplify into substantial differences among the resulting epistemic systems. The present deliberations thus bear a caution for any attempts to apply the formal work of the epistemic logician to deliberations in substantive epistemology.

The crucial point is that "epistemic logic" admits of a great variety of alternative systematizations, all of them equally "correct," given the somewhat rough-edged structure of our conception of knowledge. Epistemic logic can certainly clarify the implicit consequences of various assumptions one might want to make about the interrelationships of what we know. But it is inherently unfitted to serve us in determining the unique correctness of any one specific version of a formalized theory of knowledge.[1]

## Appendix

In this appendix we shall provide proof for two assertions made earlier in the paper:

(1) if **S** is one of the systems **T**, **S4**, **S5** then

$\lambda$ is an **S**-world $\Leftrightarrow$ i) $\lambda$ is maximal, consistent, and mp-closed
ii) $\mathbf{S}^* \subseteq \lambda$

(2) For (appropriate) systems **S** which contain the rule of epistemization we have that for any **S**-world $\lambda$

$$KA \epsilon \lambda \Leftrightarrow (\forall \kappa)\ (\kappa \text{ is an S-world and } \lambda R \kappa \Rightarrow A \epsilon \kappa)$$

The former result is stated in Lemmas 5–7, obtained *via* Lemma 2; the latter result is obtained in Lemma 9. The remaining lemmas are preparatory to these two results.

Throughout this appendix we shall make use of the following three definitions:

(I) A set $\lambda$ is **S**-*consistent* $\Leftrightarrow$
$(A_1, \ldots, A_n) \epsilon \lambda \Rightarrow \sim(A_1\ \&\ \ldots\ \&\ A_n) \notin \mathbf{S}^*]$
$\mathscr{E}(\mathbf{S}) = \{\lambda: \lambda \text{ is maximal and S-consistent}\}$

[1] This essay is a somewhat modified version of a paper of the same title written in collaboration with Arnold vander Nat and published in the *Journal of Philosophical Logic*, vol. 2 (1973).

Given an ordering of all the formulas, $A_0, A_1, A_2, \ldots, A_i, \ldots$, and given a set $\lambda$, we define a sequence of sets as follows:

(II) $f(\mathbf{S}, \lambda, 0) = \lambda$
$f(\mathbf{S}, \lambda, i+1) =$
$$= \begin{cases} f(\mathbf{S}, \lambda, i) \bigcup \{A_i\}, \text{ if } f(\mathbf{S}, \lambda, i) \bigcup \{A_i\} \text{ is } \mathbf{S}\text{-consistent} \\ f(\mathbf{S}, \lambda, i) \bigcup \{\sim A_i\}, \text{ otherwise} \end{cases}$$

(III) $\Xi(\mathbf{S}, \lambda) = \bigcup_{i \geqslant 0} f(\mathbf{S}, \lambda, i)$

*Stipulation:* In the lemmas that follow we assume a system **S** such that

(C1) $\mathbf{PC}^* \subseteq \mathbf{S}^*$
(C2) $\mathbf{S}^*$ is mp-closed

LEMMA 1. $\lambda$ is **S**-consistent$\Rightarrow \Xi(\mathbf{S}, \lambda)$ is maximal and **S**-consistent

*Proof.* Let $\lambda$ be **S**-consistent. First, $\Xi(\mathbf{S}, \lambda)$ is maximal by construction. Second, to show that $\Xi(\mathbf{S}, \lambda)$ is **S**-consistent, we show that

$\kappa$ is **S**-consistent$\Rightarrow \kappa \cup \{A\}$ is **S**-consistent or $\kappa \cup \{\sim A\}$ is **S**-consistent

Suppose not. Then $\kappa$ is **S**-consistent and $\kappa \cup \{A\}$ and $\kappa \cup \{\sim A\}$ are not. So, for some $A_1, \ldots, A_n \epsilon \kappa \cup \{A\}$ and some $B_1, \ldots, B_m \epsilon \kappa \cup \{\sim A\}$, $\sim (A_1 \,\&\, \ldots \,\&\, A_n) \,\epsilon\, \mathbf{S}^*$ and $\sim (B_1 \,\&\, \ldots \,\&\, B_m) \,\epsilon\, \mathbf{S}^*$. Since $\kappa$ is **S**-consistent, one of the $A_i$ is $A$ and one of the $B_j$ is $\sim A$, say, for convenience, $A_1$ and $B_1$ respectively. Then by (C1) and (C2), $A \supset \sim (A_2 \,\&\, \ldots \,\&\, A_n) \,\epsilon\, \mathbf{S}^*$ and $\sim A \supset \sim (B_2 \,\&\, \ldots \,\&\, B_m) \,\epsilon\, \mathbf{S}^*$, *where* $A_2 \ldots, A_n, B_2, \ldots, B_m \,\epsilon\, \kappa$. Again, by a PC-tautology $\sim (A_2 \,\&\, \ldots \,\&\, A_n \,\&\, B_2 \,\&\, \ldots \,\&\, B_m) \,\epsilon\, \mathbf{S}^*$, contradicting that $\kappa$ is **S**-consistent.

Hence, since $\lambda$ is **S**-consistent, so is each $f(\mathbf{S}, \lambda, i)$, and hence also $\Xi(\mathbf{S}, \lambda)$.

LEMMA 2. $\lambda$ is maximal and **S**-consistent$\Leftrightarrow$
$\lambda$ is maximal, consistent, mc-closed, and $\mathbf{S}^* \subseteq \lambda$

The proof of this lemma follows directly from conditions (C1) and (C2).

LEMMA 3. $A \epsilon \mathbf{S}^* \Leftrightarrow A \epsilon \bigcap \mathscr{E}(\mathbf{S})$

*Proof.* If $A \epsilon \mathbf{S}^*$ then, by Lemma 2, $A \epsilon \bigcap \mathscr{E}(\mathbf{S})$. If $A \notin \mathbf{S}^*$, then by (C1) and (C2), $\sim\sim A \notin \mathbf{S}^*$, so that $\{\sim A\}$ is **S**-consistent. By Lemma 1, $\Xi(\mathbf{S}, \{\sim A\}) \epsilon \mathscr{E}(S)$. Hence for some $\kappa \epsilon \mathscr{E}(\mathbf{S})$, $A \notin \kappa$.

LEMMA 4. $A \epsilon \mathbf{S}^* \Leftrightarrow A \epsilon \bigcap \{\lambda$: $\lambda$ is maximal, consistent, mp-closed, and $S^* \subseteq \lambda\}$

The proof of this lemma follows directly from Lemma 2 and 3.

LEMMA 5. $\lambda$ is a **T**-world$\Leftrightarrow\lambda\epsilon\mathscr{E}$(**T**)

*Proof.* If $\lambda$ is maximal and **T**-consistent, then $\lambda$ is a **T**-world since, by Lemma 2 and Lemma 4, $\mathscr{E}$(**T**) is a **T**-world structure. On the other hand, let $\lambda$ be a **T**-world. Then for some $W$, $\lambda\epsilon W$, where

$$(\forall\kappa)\ (\kappa\epsilon \vee \Rightarrow \kappa \text{ is maximal, consistent, mp-closed, } O^* \subseteq \kappa)$$
$$A\epsilon\bigcap\vee \Rightarrow KA\epsilon\bigcap W$$

We need thus only show that $\mathbf{T}^*\subseteq\lambda$. We show by induction that if $A\epsilon\mathbf{T}^*$ then $A\epsilon\bigcap W$. *Case 1*: $A$ is an axiom of **T**. Then $A\epsilon O^*$ and $A\epsilon\bigcap W$. *Case 2*: $A$ comes by *modus ponens*. Then for some $B$, $B\epsilon\mathbf{T}^*$ and $B \supset A\epsilon\mathbf{T}^*$. By inductive hypothesis, $B\epsilon\bigcap W$ and $B\supset A\epsilon\bigcap W$. Since each member of $W$ is mp-closed, $A\epsilon\bigcap W$. *Case 3*: $A$ comes by the rule of epistemization. Then, for some $B$, $A = KB$, where by inductive hypothesis, $B\epsilon\bigcap W$. But then $KB\epsilon\bigcap W$, i.e., $A\epsilon\bigcap W$.

LEMMA 6. $\lambda$ is an **S4**-world$\Leftrightarrow\lambda\epsilon\mathscr{E}$(**S4**)

LEMMA 7. $\lambda$ is an **S5**-world$\Leftrightarrow\lambda\epsilon\mathscr{E}$(**S5**)

The proofs of Lemmas 6 and 7 are entirely similar to the proof of Lemma 5.

*Stipulation:* In the lemmas that follow we assume a system **S** such that (C2) holds and such that

(C3) $\mathbf{O}^*\subseteq\mathbf{S}^*$ (so that (Cl) holds also)
(C4) $A\epsilon\mathbf{S}^*\Rightarrow KA\epsilon\mathbf{S}^*$
(C5) $\lambda$ is an **S**-world $\Leftrightarrow$ $\lambda$ is a maximal, consistent, mp-closed set such that $\mathbf{S}^*\subseteq\lambda$.

LEMMA 8. $\lambda$ is an **S**-world and $KA\notin\lambda\Rightarrow[\lambda]\bigcup\{\sim A\}$ is **S**-consistent

*Proof.* Suppose not. Then for some $A_1, \ldots, A_n\epsilon[\lambda]\bigcup\{\sim A\}$, $\sim(A_1 \,\&\ldots\&\, A_n)\epsilon\mathbf{S}^*$.

*Case 1.* None of the $A_i$ is $\sim A$. Then $A_1, \ldots, A_n\epsilon[\lambda]$, and

$\sim(A_1 \,\&\ldots\&\, A_n \,\&\, \sim A)\ \epsilon\ \mathbf{S}^*$, by (Cl) and (C2)
$A_1 \supset (A_2 \supset \ldots (A_n \supset A)\ldots)\ \epsilon\ \mathbf{S}^*$, by (Cl) and (C2)
$\kappa(A_1 \supset (A_2 \supset \ldots (A_n \supset A)\ldots))\ \epsilon\ \mathbf{S}^*$, by (C4)
$KA_1 \supset (KA_2 \supset \ldots (KA_n \supset KA)\ldots)\ \epsilon\ \mathbf{S}^*$, by (C2) and (C3)
$KA_1 \supset (KA_2 \supset \ldots (KA_n \supset KA)\ldots)\ \epsilon\ \lambda$, by (C5)

But, since $KA_1, \ldots, KA_n\epsilon\lambda$, we have $KA\epsilon\lambda$, *contra* the hypothesis.
*Case 2.* Some $A_i$ is $\sim A$, say, for convenience $A_1$. Then $A_2, \ldots, A_n\epsilon[\lambda]$, $(A_2 \,\&\ldots\&\, A_n \,\&\!\sim\! A)\epsilon\mathbf{S}^*$, and $KA_2W(KA_3\supset \ldots (KA_n\supset KA).\,.\,.)\epsilon\lambda$ as before. But, since $KA_2, \ldots, KA_n\epsilon\lambda$, $KA\epsilon\lambda$ also, contrary to hypothesis.

LEMMA 9. $\lambda$ is an **S**-world$\Rightarrow$

$$KA \in \lambda \Leftrightarrow (\forall\kappa)(\kappa \text{ is an } \mathbf{S}\text{-world and } \lambda R\kappa \Rightarrow A \in \kappa)$$

*Proof.* The sufficiency clause of Lemma 9 follows by the definition of $R$. To show the necessity clause, suppose $KA \notin \lambda$. Then by Lemma 8, $[\lambda] \cup \{\sim A\}$ is **S**-consistent. By Lemma 2 and (C5), $\Xi(\mathbf{S}, [\lambda \cup]\{\sim A\})$ is an **S**-world $\kappa$ such that $[\lambda] \subseteq \kappa$ and $A \notin \kappa$. Q.E.D.

# VII.

# Epistemic Modal Categories and the Theory of Plausibility

## 1. The Plausibility Indexing of a Propositional Set

THE idea of a measure of the "plausibility" of propositions is intended to capture, by way of a formal calculus, not the degree—familiar from probability theory—to which propositions are *likely to be true* (relative to given evidence), but the extent to which propositions *if accepted as true* would "be at home" within the wider setting of what we incline to accept. The operative issues are: does the proposition represent "exactly what one would expect"; does it "accord relatively well with our expectations"; does it "realize a rather far-out prospect, which goes quite against our expectations"; and the like. It should be noted that what is at stake here is "acceptability," and *not* how probable something is relative to what we know to be true. When I know that there are 5 books on the table, then the (relative) *probability* that there are 6 is very low indeed (viz., 0) but the plausibility of this prospect is not insubstantial (there would be nothing unusual or out of the ordinary for there to be 6 books on this table instead of 5); but that there should be 60,000 books is very implausible indeed. The plausibility of a proposition is a matter of the *depth* of the revisions in the body of our knowledge (or putative or presumptive knowledge) that would be called for *if* it were to turn out to be true (however unlikely that might be). If the propositions were, for example, to represent what we regard as a *logical* truth, then its plausibility could not be improved upon; but if it were to stipulate some very peculiar-seeming and counter-intuitive fact, it would, by contrast, be classed as highly implausible.

The aim of a plausibility indexing is to codify the comparative extent of our epistemic commitment to propositions. The informal ideas at issue here are not difficult to explain. To say that a proposition is relatively plausible is *not* to say that it is true, but only that its epistemic claims are to be viewed as relatively strong: that if it were to be true this would not surprise us, but would be something that we should welcome (from the epistemic point of view—not necessarily from others) as smoothly in accord with the general pattern of what we know. Plausibility is a sort of *potential* commitment: when we regard a statement as highly plausible we are saying that *if* we were to accept it as true, then we should be prepared to give it a very comfortable and secure place among the truths. And the more plausible the statement, the more deeply we should commit ourselves to accepting it as true if we did in fact so accept it. The allocation of plausibility-index values to a group of statements is thus a reflection of our relative degree of attachment to these statements—be it actual attachment or hypothetical attachment in the context of a certain analysis. In giving one statement a better plausibility classification than another we are saying *that if in the last resort we had to make a choice* between them, we should grant precedence to the more plausible statement.

Given a set **S** of propositions, a plausibility indexing of **S** is to consist in assigning to the elements of **S** integers from the list 0, 1, 2, ..., $n$. (No special restrictive assumptions need be made regarding the context of this propositional set **S**.) This assignment is to be designed to implement the intuitive notion that the values play the following roles in line with the idea that the smaller the plausibility index value we assign to a proposition, the greater is its plausibility:

1. 0 represents maximal plausibility (or logical certainty)
2. 1 represents high plausibility (or effective or virtual certainty)
3. 0, 1, ..., $m$ represent decreasing degrees of positive plausibility
4. $m$, $m+1$, ..., $n$ represent increasing degrees of implausibility
5. $n$ represents high or maximal implausibility.

In general, when $i<j$, then $i$ indicates a higher degree of plausibility than $j$ (and $j$ a higher degree of implausibility than $i$). The terminal value $n$ is to be large enough to permit drawing all the plausibility distinctions necessary to the case in hand.

Formally, this numerical assignment is to conform to the following rules: There is to be an indexing (called a *plausibility indexing* for the set **S**) assigning for every proposition $P \in \mathbf{S}$ an index value $/P/$ where $0 \leqslant /P/ \leqslant n$. This indexing is to be such that the following conditions hold:

(P1) Every proposition in **S** obtains a plausibility value. For every $P \in \mathbf{S}$, there is some value with $0 \leqslant k \leqslant n$ such that $/P/=k$.

(P2) Logical truths are maximally plausible; indeed they constitute the category of the maximally plausible. $\vdash P$ iff $/P/=0$.

(P3) All the propositions classed as highly plausible must be mutually compatible. The set $\{P: /P/ = 1\}$ is to be consistent.

(P4) When a certain (consistent) group of propositions entails some proposition, then this proposition cannot be less plausible than the least plausible among them. If $P_1, \ldots, P_r \vdash Q$ (with $r<1$), and $P_1, \ldots, P_r$ are mutually consistent, then $/Q/ \leqslant \max_{1 \leqslant i \leqslant r} /P_i/$.

From (P4) it follows at once that:

$$\text{If } P \vdash Q, \text{ then } /Q/ \leqslant /P/.$$

This entails that interdeducible propositions must have the same plausibility ranking. Moreover, one can readily establish the result:

$$/P \;\&\; Q/ = \max\,[/P/, /Q/].$$

The proof goes as follows:

(1) Since $P \;\&\; Q \vdash P$, we have by the preceding result that $/P/ \leqslant /P \;\&\; Q/$. And analogously, $/Q/ \leqslant /P \;\&\; Q/$. Consequently $\max\,[/P/, /Q/] \leqslant /P \;\&\; Q/$.

(2) Since $P, Q \vdash P \;\&\; Q$, we have it by (P4) that $/P \;\&\; Q/ \leqslant \max\,[/P/, /Q/]$.

(3) The desired equality relation now follows at once from (1) and (2).

Another obvious theorem (since both $P \vdash P \vee Q$ and $Q \vdash P \vee Q$) is:

$$/P \vee Q/ \leqslant \min\,[/P/, /Q/]$$

Thus in the special case that $/P/ = /Q/$ we have

$$/P \;\&\; Q/ = /P \vee Q/ = /P/ = /Q/$$

In compounding the propositions of a *fixed* plausibility level, we remain at that same level.

It is important for the applications of the ideas at issue that plausibility-indexed sets need not in general be consistent. Both $P$ and $\sim P$ could turn out to be relatively plausible. But if $P$ and $\sim P$ are both members of some plausibility-indexed set S, and $/P/$ is known, what can be inferred as to $/\sim P/$? Very little. Here we only know that:

$$\text{If } /P/ \leqslant 1, \text{ then } /\sim P/ \geqslant 2$$

This is clearly so since the family of all propositions indexed at $\leqslant 1$ must be mutually compatible. This lack of a negation-rule renders the calculus of plausibility inherently more impoverished as a mathematical formalism than the calculus of probability.

The plausibility indexes are to be so interpreted that the lower the index-number of a statement "the more deeply we would commit ourselves to accepting" this statement if we did in fact accept it. Thus the function of the indexing is to provide the machinery for an exact logical articulation of the informal idea of the "relative degree of potential commitment" to the endorsement of statements, reflecting their relative acceptability in the epistemic scheme of things.

The question of the *existence* of a plausibility-indexing of a given set **S** of propositions is, from a *formal* point of view (where issues of interpretation do not concern us) quite trivial. The following simple construction-procedure suffices to settle this issue:

(1) Whenever $P \in \mathbf{S}$ is such that $\vdash P$, then set $/P/$ at 0.
(2) Select a self-consistent $Q \in \mathbf{S}$ and for any $P \in \mathbf{S}$ set $/P/$ at 1 whenever $Q \vdash P$.
(3) Whenever $P \in \mathbf{S}$ and $/P/$ is *not* determined by the two preceding rules, then set $/P/$ at 2.

The application of this procedure will automatically issue in a plausibility-indexing that meets all the conditions specified above.

## 2. Plausibility and Probability

A plausibility-indexing is roughly analogous to an assignment of probabilistic likelihoods. But there are certain crucial and decisive exceptions, among which the following two are preeminent: (1) When statements of equal probability are conjoined, the *probability* of the resultant conjunction is (in general) diminished. But—by the above reasoning—the plausibility of the conjunction is in such cases to remain unaltered. Thus if $P_1, P_2, \ldots, P_k$ are all highly probable, their conjunction can be very improbable indeed, but if they are all very plausible—if we are "very definitely inclined" to accept each of them—we cannot in logical honesty be any the less minded to accept their conjunction. (2) When the probability of $P$ is given, that of $\sim P$ is determined, and so determines that if $P$ is rather probable (i.e., has relatively high probability), then $\sim P$ is rather improbable (i.e., has relatively low probability). By contrast, the plausibility of $\sim P$ bears no necessary relationship to that of $P$: both can together obtain relatively high or relatively low plausibility values. The fact is that plausibility is a distinct concept in its own right and is something altogether different from probability in the standard sense as explicated by the mathematical calculus of probability.

To facilitate the comparison and contrast between the probabilistic and the plausibilistic assessment of propositions, we shall tabulate the relevant fundamental rules of calculation in comparative form:

| PLAUSIBILITY ($\pi$) | PROBABILITY ($p$) |
|---|---|
| 1. The measure is to be defined over an *arbitrary* set S of propositions. Increasing $\pi$-values represent *decreasing* plausibility. | 1. The measure is to be defined over a *logically self-contained set* S of propositions that is closed under the Boolean operations of propositional combination. Increasing $p$-values represent *increasing* probability. |
| 2. $\pi(P)$ is an *integer* such that $0 \leqslant \pi(P)$ | 2. $p(P)$ is a *real number* such that $0 \leqslant p(P) \leqslant 1$ |
| 3 $\pi(P) = 0$ iff $\vdash P$ | 3. $p(P) = 1$ if $\vdash P$ |
| 4. $\pi(P \vee Q) \leqslant \min[\pi(P), \pi(Q)]$ | 4. If $\vdash \sim(P \& Q)$, then $p(P \vee Q) = p(P) + p(Q)$. And always: $\max[p(P), p(Q)] \leqslant p(P \vee Q)$ |
| 5. $\pi(P \& Q) = \max[\pi(P), \pi(Q)]$ | 5. $p(P \& Q) = p(P) \times p(Q \text{ if } P)$. And so: $p(P \& Q) \leqslant \min[p(P), p(Q)]$ |
| 6. $\pi(\sim P)$ is in general *not* functionally dependent upon $\pi(P)$.[1] | 6. $p(\sim P) = 1 - p(P)$ |
| 7. If $P_1, P_2, \ldots, P_n \vdash Q$, then $\pi(Q) \leqslant \max_i[\pi(P_i)]$ | 7. If $P_1, P_2, \ldots, P_n \vdash Q$, then there does not follow any general functional relationship between $p(Q)$ and the various $p(P_i)$. |

It might seem at first thought that one could attempt to *correlate* plausibilities and probabilities along the lines of the following rules of correlation [now writing simply $/P/$ for $\pi$ ($p$)]:

| *Plausibilistic Situation* | *Probabilistic Situation* |
|---|---|
| $/P/ = 0$ | $p(P) = 1$ |
| $/P/ = \text{n}$ (maximal) | $p(P) = 0$ |
| $0 < /P/ < \text{n}$ | $p(P) := 1 - \frac{/P/}{\text{n}}$ |

[1] It is primarily this feature that blocks a plausibility-valuation from qualifying as a many-valued logic. For further considerations regarding this issue of a negation rule see Section 5 below.

That is, one might attempt to set up a correlation of plausibilities with probabilities as being simply two different aspects of the same thing, by arranging agreement at the endpoints of the two scales with linear interpolation inbetween. But this effort rapidly comes to shipwreck, since the plausibilities of negations and conjunctions (for example) do not behave in a way accordant with probabilities. We have neither that $/{\sim}P/$ is in inevitably identical with $1 - /P/$, nor that $/P \& Q/$ is generally less than the maximum of $/P/$ and $/Q/$.

To summarize: probabilities and plausibilities are different in conception and different in effect. Even when probabilities are determined by the use of plausibility-considerations, or when plausibilities are assigned on the basis of probabilistic information, the two modes of analysis will subsequently proceed on their own separate ways. The key points of difference are:

(1) Plausibility status is preserved in conjunctions with conjuncts of similar status, but probability status not.
(2) The assignment of probability values to propositions determines those of their negations, plausibility values do not.
(3) Plausibility assignments take implicational relationships into account more extensively than probabilities do.

The sharp difference between these two modes of propositional valuation in the context of the present analysis may be highlighted as follows: consider two propositions $P$ and $Q$ and suppose that we have information about their acceptability *separately*, but seek to have information about their acceptability *jointly*. If this information is given in probabilistic terms we may find ourselves left at the starting gate. Thus if we know simply that both $P$ and $Q$ are relatively probable (say each has probability 0.4), then we know little about the probability of their conjunction (it may be 0 if $\Pr(P) = 0.45$ and $Q = {\sim}P$; again it may be 1 if $\Pr(P) = 1$ and $Q = P$). Separate probabilities do a very incomplete job of taking account of the logical relationships that emerge from conjunctions. With a probabilistic approach to selecting the propositions of a set S we cannot look at these propositions separately and seriatim, we must always look at them also conjunctively so as to take interactions into account. With plausibilities, on the other hand, it is a defensible procedure to proceed with reference to separate propositions because in the assignment of plausibility numbers the logical interrelationships among the propositions that arise in conjunction have already been taken into account. Thus, given that some set S has three elements, $P$, $Q$, $R$, that are all "very probable"—say with probabilities

3/4, 2/3, 2/3, respectively—we know little about the probability of the set as a whole. Its "axiom" $P$ & $Q$ & $R$ could have a probability as great as 8/12 or as little as 1/12 and could range from "very probable" to "rather improbable." But in terms of plausibilities, if we know that all three elements are "very plausible" then we know that the set as a whole—or rather its "axiom" $P$ & $Q$ & $R$—must also be very plausible.

## 3. Motivation and Interpretation

One of the best ways of conceptualizing a plausibility indexing is to think of inferences made from a group of premisses belonging to very different epistemic categories, ordered in point of solidity and security. On the principle that "a chain is no stronger than its weakest link," the status of a conclusion will clearly be determined by that of its "weakest" premiss. The old principle of modal logic, that the conclusion follows the weakest premiss (*sequitur conclusio peiorem partem*) becomes applicable in the present case also.

The pivotal fact is that the usual rule of probabilistic degradation in conjunctions—according to which Pr($P$ & $Q$) is in general substantially less than Pr($P$) or Pr($Q$)—envisages the case of conjoining claims that function in an essentially separate and discrete way. By contrast, the plausibilistic principle of conjunction envisages the case of *systematic interdependence*, where the acceptability of one component is part and parcel with that of the whole to which it belongs. Plausibilistic combination is not modelled on the aggregation of discrete units, but on that of the accession of an entire systematic whole, with a view to the acceptance not of this or that separate item, but of a whole network of interlocked and interdependent parts. Here the whole rests on the same footing as its parts—they stand or fall together, and have a common and *shared* status of credibility. Indeed, in such cases it is to be precisely its membership in a certain systematic whole that is to be determinative of the epistemic status of the parts. The status of a large-scale scientific theory or discipline provides a guiding analogy—a system of physical geometry, for example, must be accepted *en bloc* as a whole, its acceptability cannot be motivated as a compilation of discrete and separately confirmed bits and pieces. (Recall the teachings of Pierre Duhem on this head.) Thus quite different paradigms are at issue in the plausibilistic and probabilistic spheres. The probabilistic mode of combination addresses itself to the essentially *aggregative* case whose model is the distributive acceptance of discrete items, whereas the plausibilistic mode of combination addresses itself to the essentially *systematic* case whose model is the collective acceptance of a unified whole, each of whose components is effectively interlocked with the rest.

In accordance with this line of thought, let us suppose as given a formal system of some sort, based on certain axioms and rules of inference which provide a corresponding notion of a theoremhood $\vdash P$. Consider now some sequence of further nontheorematic, mutually compatible, accepted statements: $S_1^1, S_2^1, \ldots, S_{n_1}^1, S_1^2, S_2^2, \ldots, S_{n_2}^2, S_1^k, S_2^k, \ldots, S_{n_k}^k$. Now we construct the plausibility categories $\mathbf{C}_i$ as follows:

(0) $\mathbf{C}_0$ is the set of all $P$ such that $\vdash P$.

(1) $\mathbf{C}_1$ is the set of all $P$ such that $S_1^1, \ldots, S_{n_1}^1 \vdash P$.

(2) $\mathbf{C}_2$ is the set of all $P$ such that $S_1^1, \ldots, S_{n_1}^1, S_1^2, \ldots, S_{n_2}^2 \vdash P$.
*Et cetera* until

(k) $\mathbf{C}_k$ is the set of all $P$ such that $S_1^1, \ldots, S_{n_k}^k \vdash P$.

We shall suppose that when plausibility categories are given in this way they are nonredundant, i.e., that every set $\mathbf{C}_i$ includes some statements not contained in $\mathbf{C}_{i-1}$. We suppose too an "unwritten" (k+1)st modal category as a "catch all" for all statements of the propositional set **S** in view that are not otherwise accommodated when $\mathbf{C}_i$ membership is used to guide the establishment of a plausibility indexing of **S** via the rule that when $P \in \mathbf{S}$, then $/P/ = i$ iff $P \in \mathbf{C}_i$.

It is readily seen that, given this construction of the set $\mathbf{C}_i$, all the conditions for a plausibility indexing are at once satisfied.

## 4. Some Sample Applications

The workings of the preceding ideas can be clarified by examining some illustrative examples. Let us begin with a schematic example. Consider the propositional set:

$$\mathbf{S} = \{p, \sim p, q, \sim q, q \supset r, p \supset \sim r\}$$

It is readily seen that the following assignment provides a plausibility-index for this set:

| | *Proposition* | *Index* |
|---|---|---|
| (1) | p | 4 |
| (2) | ~p | 3 |
| (3) | q | 3 |
| (4) | ~q | 2 |
| (5) | q⊃r | 2 |
| (6) | p⊃~r | 3 |

In checking the adequacy of this indexing, the following implication-relations must be heeded:

| | |
|---|---|
| (2) ⊢ (6) | 3 ⊢ 3 |
| (4) ⊢ (5) | 2 ⊢ 2 |
| (3), (5), (6) ⊢ (2) | 3, 2, 3 ⊢ 3 |
| (1), (5), (6) ⊢ (4) | 4, 2, 3 ⊢ 2 |

These relationships serve to show that the specified value-assignment provides a plausibility-indexing.

The preceding example, helpful though it is in illustrating the generic structure of the analysis, is still abstract and schematic. It is desirable to indicate some applications of a more concrete and content-laden sort. Consider the following set **S** of accepted beliefs, assumed to correspond to what is in fact the case:

$p_1$ = This coin is a penny.
$p_2$ = This coin is not a dime.
$p_3$ = This coin is made of[2] copper.
$p_4$ = This coin is not made of silver.
$p_5$ = All pennies are made of copper.[3]
$p_6$ = All dimes are made of silver.

We are now to introduce the counterfactual hypothesis $\sim p_2$:

Assume that this coin were a dime.

The propositional set **S′** that results from replacing $p_2$ by $\sim p_2$ is clearly still inconsistent. It has two maximal consistent subsets (m.c.s.) that are not incompatible with $\sim p_2$:[4]

$\mathbf{S}'_1 = \{\sim p_2, p_3, p_4, p_5\}$
$\mathbf{S}'_2 = \{\sim p_2, p_5, p_6\}$

In the absence of any further plausibilistic basis for choosing between $\mathbf{S}'_1$ and $\mathbf{S}'_2$, we arrive at the following axiomatization of the "inevitable-consequences" of the hypothesis (viz., those theses which follow from *all* of the eligible m.c.s.):

$$\sim p_2 \& p_5 \& ([p_3 \& p_4] \vee p_6)$$

[2] Through this example, "is made of" is to be construed as "is made *predominantly* of" rather than "is made *in part* of".

[3] Through this example "penny" is to be understood as "penny *in current circulation*." The fact that in some past periods pennies were made of other metals is thus immaterial.

[4] We suppose that $p_1$ is inconsistent with $p_2$, so that $p_1$ will occur in no $p_2$-containing m.c.s.

We would thus derive, *inter alia*, the result $p_3 \vee p_6$, and so, thanks to the incompatibility-relations at issue, we could, for example, obtain: also $\sim p_3 \vee \sim p_6$. This leads to the counterfactual:[5]

> If this penny were a dime then either it would not be made of copper (since dimes are made of silver) or all dimes would not be made of silver (since this coin is made of copper, not silver).

However, if more specific preferential information were available, we could go beyond this result in definiteness. We might, for example, want to implement the standpoint of a differential treatment of laws and factual theses as regards plausibility. In terms of plausibility indexing, two alternative policies offer themselves as to the relative position of facts (as presented in particular statements) and laws (as presented as universal generalizations):

*Alternative 1*: Particular statements are to be more plausible (and so receive a lower index value) than generalizations.

*Alternative 2*: Generalizations are to be more plausible (and so receive a lower index value) than particular statements.

On the first alternative we should obtain:

1-indexed propositions: $\sim p_2$ (the hypothesis at issue)
2-indexed propositions: $p_3, p_4$
3-indexed propositions: $p_5, p_6$

This plausibility indexing results in the preferential selection of the particular m.c.s. $S'_1$. Correspondingly, we should give up not only $p_1$ and $p_2$ (which follows as a matter of course), but also $p_6$. We should accordingly establish the counterfactual:

> If this penny were a dime, then all dimes would not be made of silver (since this penny is made of copper).

Note that—as our plausibility-assessing procedure insists—we have made a generalization give way to a particular statement.

By contrast, if we adopted Alternative 2, we should obtain:

1-indexed propositions: $\sim p_2$
2-indexed propositions: $p_5, p_6$
3-indexed propositions: $p_3, p_4$

This plausibility-indexing results in the preferential selection of the particular m.c.s. $S'_2$. Correspondingly, $p_1$ and $p_2$ are joined in banishment by $p_3$ and $p_4$. We accordingly establish the counterfactual:

[5] For this approach to the analysis of counterfactual conditionals *via* their associated maximal consistent subsets, see the author's *Hypothetical Reasoning* (Amsterdam, 1964).

If this penny were a dime, then it would be made of silver and not of copper (since dimes are made of silver).

Here—as expected—we have made particular statements give way to generalizations.

Let us consider yet another example of a somewhat different type. Suppose we interrogate four sources $X_1$–$X_4$ with respect to two propositions that interest us ($p$ and $q$) and find them to make the following declarations:

$X_1$: $p \supset q$
$X_2$: $p \vee q$
$X_3$: $p \,\&\, {\sim}q$
$X_4$: $q$

The sources might, moreover, afford us a self-appraisal of their relative expertise in the topical area at issue, or perhaps we can ourselves obtain an independent evaluation. Let the results be as follows:

$X_1$ is very knowledgeable.
$X_2$ is very knowledgeable.
$X_3$ is reasonably knowledgeable.
$X_4$ is relatively uninformed.

To arrive at a consensus position, we could apply the plausibility machinery in the familiar way. We begin with the data:

$$\mathbf{S} = \{p \supset q, p \vee q, p \,\&\, {\sim}q, q\}.$$

The m.c.s. of **S** are

$\mathbf{S}_1 = \{p \supset q, p \vee q, q\}$ axiomatized by $q$.
$\mathbf{S}_2 = \{p \vee q, p \,\&\, {\sim}q\}$ axiomatized by $p \,\&\, {\sim}q$.

To develop a plausibility index we could begin with the correlation

| *Expertise* | *Plausibility* |
|---|---|
| highly expert | 2 |
| very knowledgeable | 3 |
| reasonably knowledgeable | 4 |
| relatively uninformed | 5 |
| utterly ignorant | 6 |

Thus we obtain the indexing

$$/p \supset q/ = 3$$
$$/p \vee q/ = 3$$
$$/p \,\&\, {\sim}q/ = 4$$
$$/q/ = 5$$

This assignment in fact provides a full-fledged plausibility index.

The plausibility situation for the m.c.s. is now as follows:

| | *Maximum* | *Average* |
|---|---|---|
| $S_1$ (3, 3, 5) | 5 | 3.7 |
| $S_2$ (3, 4) | 4 | 3.5 |

Accordingly, the preferred m.c.s. is $S_2$ which is axiomatized by $p$ & $\sim q$. This proposition could thus serve us as the "consensus result" of the initial declarations. (Note that the upshot is a preference for $X_3$'s report to that of the more knowledgable $X_1$.)

It should be noted that if we treated the sources of this example as of uniform knowledgeability then we should have no basis of preference among the m.c.s. In this case, one would arrive simply at the outcome $p \vee q$ to represent the (now substantially more meager) consensus result—unless some other criterion of m.c.s. preference were adopted.

## 5. A Negation Rule in Special Cases

A plausibility indexing of a set **S** may cease to be such when further elements are added to **S**, even when those elements are no more than mere consequences of **S**-elements. Thus consider

$$\mathbf{S} = \{p, \sim p, \sim r)$$

The assignment

$$/p/ = 2$$
$$/\sim p/ = 2$$
$$/\sim r/ = 3$$

provides a (possible) plausibility indexing of this set. But if we change **S** to **S′** by adding $p \vee r$ (a mere consequence of the **S**-element $p$), then we have

$$\mathbf{S}' = \{p, \sim p, p \vee r, \sim r\}$$

And now we can have no indexing of the form

$$/p/ = 2$$
$$/\sim p/ = 2$$
$$/\sim r/ = 3$$
$$/p \vee \sim r/ = x$$

For since

$$p \vdash p \vee \sim r \quad \text{we have } x = /p \vee \sim r/ \leqslant /p/ = 2$$

And since

$$\sim p, p \vee \sim r \vdash \sim r \quad \text{we have } /\sim r/ \leqslant \max\,[2, x] = 2$$

And this finding contradicts the specified indexing assignment $/\sim r/ = 3$. We now cannot continue to accept the original indexing.

Thus we cannot in general assume that the plausibility-indexing of a set will remain workable in the face of a supplementation of this set by some of its own logical consequences.

But let us suppose that we have to do with the special case in which a set **S** is *inferentially closed*, in the sense of containing all of the logical consequences that can be drawn from sets of (mutually compatible) premisses contained in **S**. Then, in this special circumstance, various significant consequences ensue.

*Theorem 1*: Let **S** be an inferentially closed, plausibility-indexed set of propositions which contains an inconsistent m-ad; that is, **S** contains a minimally inconsistent group of propositions $P_1, P_2, \ldots, P_m$; and thus $P_2, P_3, \ldots, P_m \vdash \sim P_1$. Then no self-consistent proposition $Q \in \mathbf{S}$ can possibly fare worse in plausibility than the worst off of the $P_i$. That is, if we let

$$\theta = \max[/P_1/, /P_2/, \ldots, /P_m/]$$

then no proposition can have a plausibility value $> \theta$.

PROOF. Consider an arbitrary selfconsistent proposition $Q$. Then $P_1 \vdash P_1 \vee Q$ so that $/P_1 \vee Q/ \leqslant /P_1/ \leqslant \theta$.[6] And furthermore $P_2, P_3, \ldots, P_m, P_1 \vee Q \vdash Q$ because $P_2, P_3, \ldots P_m \vdash \sim P_1$ and $\sim P_1, P_1 \vee Q \vdash Q$, so that

$$/Q/ \leqslant \max[/P_2/, /P_3/, \ldots, /P_m/, /P_1 \vee Q/] \leqslant \theta \text{ Q.E.D.}$$

*Theorem* 2: Let the plausibility-indexed set **S** be an inferentially closed set of propositions that contains at least two inconsistent j-ads, $P_1, P_2, \ldots, P_r$ and $Q_1, Q_2, \ldots, Q_s$. Then the plausibilistically worst-faring $P_i$ will be exactly the same in plausibility status as the most implausible $Q_j$.

PROOF. Let

$$\theta_1 = \max[/P_1, /P_2/, \ldots, /P_r/]$$
$$\theta_2 = \max[/Q_1, /Q_2/, \ldots, /Q_s/]$$

By Theorem 1,

$$/Q_1/ \leqslant \theta_1, /Q_2/ \leqslant \theta_1, \ldots, /Q_n/ \leqslant \theta_1$$

Thus

$$\theta_2 \leqslant \theta_1$$

By parity of reasoning we can also establish the reverse inequality:

$$\theta_1 \leqslant \theta_2$$

[6] In general, of course, $/P_1 \vee Q/$ need not be defined, since $P_1 \vee Q$ may fail to be an element of **S**, and so this statement may lack a plausibility value assignment. It is at just this point that the supposition of inferential closure becomes crucial.

Therefore

$\theta_1 = \theta_2$. Q.E.D

It follows that all inconsistent j-ads lead to a unique value which is altogether maximal in the sense that no (selfconsistent) **S**-element can possibly take on a greater plausibility value. Thus we might as well set $n$ at this (unique) worst value for the inconsistent j-ads, it being pointless to set $n$ at some value larger than we need to for accommodating the element of **S**.[7]

We now obtain the result that

*Theorem* 3: If the plausibilistically indexed set **S** be an inferentially closed set of propositions that contain some inconsistent j-ad ($P_1, P_2, \ldots, P_k$), then some member of this set of $P_i$ will take on an index value of $n$.

PROOF. This results immediately from Theorem 2.

It follows at once as a corollary that if **S** is an inferentially closed, plausibility-indexed set of propositions that contains an inconsistent dyad, i.e., a pair of separately selfconsistent but mutually contradictory propositions $P$, $\sim P$, then one member of this pair must assume the index value $n$, i.e.,

Either $/P/ = n$ or $/\sim P/ = n$.

We thus obtain at least a partial negation rule for inferentially closed sets:

If $/P/ \neq n$, and $/\sim P/$ is defined (i.e., $\sim P \epsilon \mathbf{S}$), then $/\sim P/ = n$.

Accordingly, whenever $/P/$ is different from $n$, then $/\sim P/$ must be $n$. (Since every tautology has a plausibility-value of 0, and so one that is $\neq n$, it follows at once that every contradition belonging to a set **S** of the sort in question must assume a plausibility value of $n$.)

However, we remain in ignorance as to the value of $/\sim P/$ when we are given that $/P/ = n$, so that even this special negation rule is only a partial one. Specifically, there is no systematic reason why we could not have it that both $/P/ = n$ and $/\sim P/ = n$. Thus given simply that $/P/ = n$, $/\sim P/$ can range over the entire spectrum from 0 to $n$, and remains altogether indeterminate. But apart from the special class of maximally implausible (i.e., $n$-obtaining) propositions, the rule always determines the plausibility status of negations pro-

[7] Alternatively, one could reserve $n$ for contradictions, and let some $m < n$ be the worst possible plausibility for contingent propositions. But for reasons of simplicity and economy of apparatus one might as well let $n$ do double duty for both.

vided that the fundamental—and somewhat far-fetched—assumption is satisfied that we are dealing with an inferentially closed propositional set.

## 6. Relation to an Earlier Treatment

The system of *modal categories* as explained by the author in a previous publication[8] is given by the following rules (where $P$, $Q$, $R$ range over propositions, and $\mathbf{M}_0$, $\mathbf{M}_1$, ..., $\mathbf{M}_n$ over propositional sets designated as "modal categories"):

(C1) $\vdash P$ iff $P \in \mathbf{M}_0$

(C2) If $i \leqslant j$, then $\mathbf{M}_i \subseteq \mathbf{M}_j$

(C3) $\mathbf{M}_1$ is consistent (i.e., all the propositions of $\mathbf{M}_1$ are mutually compatible)

(C4) If $P \vdash Q$ and $P \in \mathbf{M}_i$, then $Q \in \mathbf{M}_i$

(C5) $P \in \mathbf{M}_k$ for some $0 \leqslant k \leqslant n$.

A system of modal categories is also *conjunctive* (or "closed under conjunction" as it was put in *Hypothetical Reasoning*), when it satisfies the further condition

(C6) If $P \in \mathbf{M}_i$ and $Q \in \mathbf{M}_i$, then $(P \& Q) \in \mathbf{M}_i$.

When this condition is also met, then it can be shown that the categorizing system is in effect a plausibility indexing. To show this, we adopt the rule that the plausibility index of any proposition P is to be obtained by the principle:

$/P/ = k$ iff $P \in \mathbf{M}_k$ and $\mathbf{M}_k$ is the minimally indexed modal category containing $P$.

Intuitively, we are to think of the $\mathbf{M}_i$ as a cumulatively expanding series of sets, where each one in the series includes all its predecessors, and where a proposition with the plausibility index value k makes its first appearance in the k-th member of this series.

Now, returning to the plausibility principles enumerated on pp. 116–17 above, we observe as follows:

(1) (P1) is guaranteed by (C5)

(2) (P2) is guaranteed by (C1)

(3) (P3) is guaranteed by (C3)

(4) So we have only to show that (P4) holds in the system.

Suppose $P_1, \ldots, P_m \vdash Q$ where $P_1, \ldots, P_m$ are consistent. Then by (C5) each of the $P_i$, $1 \leqslant i \leqslant m$, is in some $\mathbf{M}_j$, $0 \leqslant j \leqslant n$, which by (C2)

*Hypothetical Reasoning*, *op. cit.*

yields that there is a j, $0 \leqslant j \leqslant n$, such that for all the $P_i$, $1 \leqslant i \leqslant n$, $P_i \epsilon \mathbf{M}_j$ and therefore $\max_{1 \leqslant i \leqslant m} /P_i/ \geqslant j$. Hence by (C6)

$$P_1 \ \& \ \ldots \ \& \ P_m \epsilon \mathbf{M}_j.$$

and since $P_1, \ldots, P_m \vdash Q$ this yields by (C4) that $Q \epsilon \mathbf{M}_j$, which means that $/Q/ \leqslant j$ which yields that $/Q/ \leqslant \max_{1 \leqslant i \leqslant m} /P_i/$. Q.E.D.

Thus, a conjunctive modal categorization yields a plausibility indexing. Conversely, it can also be shown that the preceding rule of correspondence leads to the upshot that any plausibility indexing generates a family of modal categories.

PROOF: We adopt the rule that the modal category membership of any proposition $P$ is to be obtained by the principle:

$$P \epsilon \mathbf{M}_k \text{ whenever } /P/ \leqslant k.$$

Then clearly:

(1) (C1) is guaranteed by (P2)
(2) (C2) is guaranteed by the basic rule
(3) (C3) is guaranteed by (P3)
(4) (C4) is guaranteed by (P4)
(5) (C5) is guaranteed by (P1)
(6) (C6) is guaranteed by the rule:

$$/P \ \& \ Q/ = \max [/P/, /Q/].$$

It follows that the procedure of plausibility indexing is essentially equivalent to that of modal categories in the special case of categories closed under conjunction.

For further elaboration of the concept of modal categories the reader may refer to *Hypothetical Reasoning* (Amsterdam, 1964).

## 7. Hamblin's Concept of "Plausibility" and Shackle's "Potential Surprise"

Our specified concept of *plausibility* is different from—though not entirely unrelated to—another version of "plausibility" proposed in an article by C. L. Hamblin.[9] Hamblin's plausibility measure is defined on the range from 0 to 1 (*this* difference from ours is immaterial) subject to three conditions:

(1) At least one state description (s.d.) obtains the maximal index value (i.e., 1).
(2) If $P$ is "known to be false," its index value must be the minimal index value (i.e., 0).
(3) The index value of a disjunction is the largest index value of its disjuncts.

[9] "The Modal 'Probably'," *Mind*, vol. 68 (1959), pp. 234–240.

Given these rules, the plausibility index of any proposition is determined as the maximum index value of the s.d. accordant with it. A proposition can be said to be plausible *per se* if its index value exceeds some predesignated threshold quantity. As expected, a conjunction of plausible alternatives need not itself be plausible—it may indeed be inconsistent.

Neither of Hamblin's conditions (1) or (3) need hold for our procedure for plausibility indexing. Nor must this be viewed as unfortunate. Why should one of the ultimate alternatives (i.e., s.d.) have to be "maximally plausible" or any intuitive construction of this idea? On the other hand, on any intuitive view of the matter it can certainly happen for some proposition $P$, that *both* $P$ *and* $\sim P$ should fare in only a fair-to-middling way as to "plausibility". (For example: "On the next toss this die will come up $\leqslant 3$," "On the next toss this die will come up $>3$.") But the proposition $P \vee \sim P$ will certainly and inevitably have to have maximal plausibility; and thus one (and by symmetry both) of the disjuncts $P$, $\sim P$ would by Hamblin's rule (3) have to be maximally plausible.

This last point can be transformed into a rather fundamental objection to the claims of Hamblin's concept to qualify as a conception of "plausibility" in a sense consonant with our common understanding of this idea. On any natural interpretation of this conception, it should certainly be possible—at least in some cases—to divide the spectrum of possibilities into a family of alternatives $A_1, A_2, \ldots, A_n$ such that (1) all the $A_i$ are alike in point of plausibility, and (2) the $A_i$ are not all maximally plausible. In such a case all the alternatives, individually considered, are alike relatively implausible. It cannot but count as a flaw in Hamblin's conception that it excludes this possibility as a matter of principle. This flaw seems to me to render Hamblin's construction of "plausibility" an implausible one. Hamblin rightly insists that each of several (mutually exclusive) alternatives might be plausible; he goes astray in ruling out the cognate possibility that each one of several (mutually exhaustive) alternatives be implausible.

Hamblin proposes to define a (non-standard) notion of probability —let $\Pi(P)$ mean "$P$ is H-probable" (probable in Hamblin's sense)— as follows:

$$\Pi(P) \text{ iff } P \text{ is more plausible than } \sim P.$$

Hamblin rightly notes that—given his concept of plausibility—*this* (distinctly non-standard) concept of probability will be subject to the rule

$$(\Pi) \quad \text{If } \Pi(P) \text{ and } \Pi(Q), \text{ then } \Pi(P \ \& \ Q)$$

unlike the situation in the case of the orthodox concept of probability. Now if we were to adopt Hamblin's definition for $\Pi(P)$ with respect to *our* concept of plausibility, then we should not be able to obtain the thesis ($\Pi$). For then we should have

$$\Pi(P) \text{ iff } /P/ < /{\sim}P/$$

Thus ($\Pi$) now becomes

$$(\Pi') \quad \text{If } /P/ < /{\sim}P/ \text{ and } /Q/ < /{\sim}Q/,$$

then

$$/P \,\&\, Q/ < /{\sim}(P \,\&\, Q)/ \text{ (or equivalently, } < /{\sim}P \vee {\sim}Q/)$$

But this is a plausibility-indexing principle that cannot be obtained from our rules without the substantial addition of further special and restrictive assumptions

However, let us introduce a concept of "improbability," analogous to Hamblin's "probability" by the analogous definition:

$$\bar{\Pi}(P) \text{ iff } \bar{\Pi}({\sim}P) \text{ iff } {\sim}P \text{ is more plausible than } P$$

We can now establish—with respect to our own concept of plausibility—that the following ($\Pi$)-analogue obtains:

$$(\bar{\Pi}) \quad \text{If } \bar{\Pi}(P) \text{ and } \bar{\Pi}(Q), \text{ then } \bar{\Pi}(P \,\&\, Q).$$

PROOF:

| | |
|---|---|
| 1. $/{\sim}P/ < /P/$ | by hypothesis $\bar{\Pi}(P)$ |
| 2. $/{\sim}Q/ < /Q/$ | by hypothesis $\bar{\Pi}(Q)$ |
| 3. $\max[/P/, /Q/] \leqslant /P \,\&\, Q/$ | by the entailments |
| 4. $\max[/{\sim}P/, /{\sim}Q/] < /P \,\&\, Q/$ | by (1)–(3) |
| 5. $/{\sim}P \vee {\sim}Q/ \leqslant \min[/{\sim}P/, /{\sim}Q/]$ | by the entailments |
| 6. $/{\sim}P \vee {\sim}Q/ < /P \,\&\, Q/$ | by (4), (5) |
| 7. $\bar{\Pi}(P \,\&\, Q)$ | by (6). |

Thus, interestingly enough, it is our concept of "improbability" (rather than "probability") that satisfies the characteristic condition for H-probability.

Hamblin offers the intriguing suggestion that his (non-probabilistic) concept of plausibility can be used to explicate G. L. S. Shackle's idea of *potential surprise*.[10] The "surprise" at issue arises when something "improbable" turns out to be true. Shackle most emphatically stresses that the "probability" here at issue is not the orthodox one, and Hamblin suggests that his concept of plausibility provides an adequate construction of Shackle's idea. But this is surely incorrect. There is nothing whatever in Shackle's discussion

[10] See Shackle's *Uncertainty in Economics* (Cambridge, 1955), especially Chapter II.

to warrant the Hamblinesque principle that one among mutually exclusive alternatives must be accorded a minimal value of potential surprise (=maximal plausibility). Indeed this seems quite counter to the whole tendency of Shackle's conception.

It would seem, however, that a concept defined in terms of our proposed style of plausibilities—via the stipulation that the potential surprise of a proposition is simply to stand in a reverse or inverse relationship to its plausibility—would answer quite well to the tendency of Shackle's ideas. Accordingly, the relative "potential surprise" of a particular truth-finding with respect to a given plausibility assignment could be assessed in terms of the relative plausibility/implausibility of the proposition in question, so that we would be quite surprised to find an implausible proposition to be true and not surprised if a plausible one is. An example will serve to clarify this.

Let us begin with the following set of data:

$$\mathbf{S} = \{p, p \supset q, \sim q, q \supset r, r\}.$$

We may suppose the following plausibility indexing of **S**:

$$/p/ = 2$$
$$/p \supset q/ = 3$$
$$/\sim q/ = 4$$
$$/q \supset r/ = 4$$
$$/r/ = 4$$

A plausibility survey of the m.c.s. yields the following results:

| *m.c.s.* | *plausibility* | *average plausibility* |
|---|---|---|
| $\{p, p \supset q, q \supset r, r\}$ | (2, 3, 4, 4) | 13/4 |
| $\{p, \sim q, q \supset r, r)$ | (2, 4, 4, 4) | 14/4 |
| $(\{p \supset q, \sim q, q \supset r, r\}$ | (3, 4, 4, 4) | 15/4 |

No matter how we resolve the issue of m.c.s. eligibility here, we shall obtain the inevitable consequences that follow from all of the m.c.s. alike. Note that these will include both $q \supset r$ and $r$, despite the relatively high plausibility-index values of these propositions, indicative of their initial implausibility. And the "potential surprise" of these findings is relatively high precisely because we are initially minded to regard them as relatively implausible.

As this example suggests, the plausibility-index value of a proposition may be regarded, from another angle, as a measure of the "potential surprise" that ensues when we find it to be true. The lower its index value, the less the surprise potential; with propositions indexed at (say) less than 2 we should find no occasion for surprise at all; truth-finding "would be only what we expect." But the larger the index value, we (i.e., whoever provides the plausibility indexing) have increasing occasion for surprise; the more implausible we had deemed the proposition when setting up the preferential criterion $\mathscr{P}$, the more appropriate is "surprise" at finding that the analysis based upon $\mathscr{P}$ yields this proposition as a result. We would accordingly be motivated to adopt the rule that the plausibility-index value of a proposition is—from a variant point of view—simply identifiable with its potential for surprising us, should it eventuate as true. From this standpoint, our theory of plausibility might well provide the desideratum (stressed by Hamblin) of a formalization of Shackle's conception of "potential surprise."

## 8. Conclusion

The discussion has sought to present the formal machinery for a measure of the plausibility of a family of theses. The prime interest of such a theory lies in its epistemological applications. These relate primarily to situations of conflict, when the "givens" we have in hand conflict with one another. Probability does not help us here: we cannot compute probabilities relative to an inconsistent information base.[11] It is just in cases of this sort, however, that plausibilistic considerations are intended to come into their own. For they are designed precisely to handle issues of this sort, given that the motivating idea is that the relative plausibility of a proposition is to reflect specifically the extent to which we are inclined to grant it precedence in the event of a conflict with other propositions that we are also minded to accept. It is in the sense of these applications that the theory of plausibility is to afford a framework for articulating what we have characterized as a system of epistemically based modal categories.[12]

[11] $\Pr(P/Q_1 \,\&\, Q_2 \,\&\, \ldots \,\&\, Q_n) = [\Pr(P \,\&\, Q_1 \,\&\, Q_2 \,\&\, \ldots \,\&\, Q_n)/\Pr(Q_1 \,\&\, Q_2 \,\&\, \ldots \,\&\, Q_n)]$, which is undefined when the $Q_i$ are mutually inconsistent.

[12] For a detailed consideration of various applications of such plausibility categorizations see the writer's book on *The Coherence Theory of Truth* (Oxford, 1973), from which parts of the present discussion have been abstracted.

# VIII.

# Restricted Inference and Inferential Myopia in Epistemic Logic

## 1. The Concept of Restricted Proof and Restricted Consequence

GIVEN a system **L** of deductive logic, it seems natural that one should be interested not only in the question of what can be demonstrated but also in that of what can be demonstrated *simply*. The idea of a *simple* proof or demonstration as such is, of course, very vague and imprecise. We propose to introduce some formal machinery for its precise articulation.

Let **L** be a system of natural deduction based on the rules $R_1$, $R_2$, . . ., $R_n$. We may now introduce the following definitions:

1. Let $S = \langle m_1, m_2, \ldots, m_n \rangle$ be a sequence of non-negative integers, each correlated with one of the $R_i$ and let $k$ be any integer greater than or equal to 2. Then an *(S, k)-restricted proof* of the proposition $B$ from the premisses $A_1, \ldots, A_n$ is one that (i) applies any rule $R_i$ no more than $m_i$ times, and (ii) is no longer than $k$ lines in length (so $2 \leqslant k \leqslant n + m_1 + m_2 + \ldots + m_n$). We shall write

$$A_1, \ldots, A_n \,[S, k \vdash B$$

to indicate that $B$ is $(S, k)$-restrictedly provable from the premisses $A_1, \ldots, A_n$.[1]

2. $B$ is an *(S, k)-restricted consequence* of the premisses $A_1, \ldots,$ $A_n$ if there exists a restricted proof that moves to $B$ as conclusion from $A_1, \ldots, A_n$ as premisses in no more than $k$ steps, i.e., when

$$A_1, \ldots, A_n \,[S, k+n \vdash B$$

In the special case that $k$ is so fixed that no restriction over and above that inherent in $S$ itself is imposed—i.e., when $k = m_1 + m_2 + \ldots + m_n$—we shall speak simply of $S$-restricted proofs and consequences. In the special where the set $S = \langle m_1, \ldots, m_n \rangle$ is such

[1] Note that the inference-claim is automatically false when $k > n$.

that all the $m_i$ are equal, $j = m_1 = \ldots = m_n$, we shall speak simply of $(j, k)$-*restricted* proofs and consequences.

When the above two special cases are combined—i.e., when the proof of $B$ from $A_1 \ldots A_n$ is no longer than $k$ lines *in toto* or when the inference to $B$ from the $A_i$ proceeds in no more than $k$ steps—we may speak simply of a *k-restricted proof* or *consequence*, respectively.

The relationships inherent in the above may be summarized as follows:

(1) $A_1, \ldots, A_n[S, k \vdash B$ This is basic

(1′) $A_1, \ldots, A_n(S, k \vdash B =_{Df} A_1, \ldots, A_n[S, n+k \vdash B$

(2) $A_1, \ldots, A_n[S \vdash B =_{Df} A_1, \ldots, A_n[S, m_1 + \ldots + m_n \vdash B$

(2′) $A_1, \ldots, A_n(S \vdash B =_{Df} A_1, \ldots, A_n[S, n+m_1+ \ldots + m_n \vdash B$

(3) $A_1, \ldots, A_n[j, k \vdash B =_{Df} A_1, \ldots, A_n[\langle j\ j \ldots j\rangle, k \vdash B$

(3′) $A_1, \ldots, A_n(j, k \vdash B =_{Df} A_1, \ldots, A_n[(\langle j, j, \ldots, j\rangle, n+k \vdash B]$

(4) $A_1, \ldots, A_n[k \vdash B =_{Df} A_1, \ldots, A_n[\langle k, k, \ldots, k\rangle, k \vdash B$

(4′) $A_1, \ldots, A_n(k \vdash B =_{Df} A_1, \ldots, A_n[\langle k, k, \ldots, k\rangle, n+k \vdash B$

The following remarks are in order:

1. There is a uniform modification (viz., the addition of $n$) in going from an unprimed relationship to its prime-indexed counterpart.
2. This relationship is such that in the primed version the number of premisses is immaterial: it is only the number of inferential steps beyond the premisses that is here at issue.
3. In cases (4)–(4′) it is only over-all length of the proof or consequence-deduction that matters. All other cases contemplate a more restrictive limitation on the number of times a given rule can be used.
4. Cases (1)–(1′) and (2)–(2′) are prepared to treat the rules of inference differentially. The other cases treat all rules alike.
5. Cases (1)–(1′) and (2)–(2′) are prepared to let $m_j$ be $0$, thus in effect proscribing the use of certain rules in drawing inferences.
6. Although none of these notions of inference allows unrestricted application of a rule (the $m_i$ are integers), this case can be covered in either of two ways—(i) by letting the $m_i$ be either integers or $\aleph_0$ or (ii) by taking the existential quantification of the relevant notion of inference with respect to the $m_i$ in question.

The idea underlying this articulation of various modes of restricted proof (or consequence) is a restriction in the number of times that the several rules of inference can be employed in a proof—and correspondingly a restriction in the over-all length of the proof itself. When that restriction is tight enough, we obtain a correspondingly restrictive notion of simplicity of proof.

One obvious application of this machinery is as follows: Given two systems **L** and **L′** such that every rule of **L** is a rule of **L′**, but not conversely, any inference relation for **L** expressible in terms of the preceding concepts will be identical with some inference relation for **L′**. If $S = \langle m_1, \ldots, m_n \rangle$ and $S' = \langle n_l, \ldots, n_k \rangle$, where each of the $m_1$ is correlated with a rule $R_i$ of **L** and each of the $n_j$ is correlated with a rule $R'_j$ of **L′**, and $n_j = m_i$ if $R'_j$ is $R_i$ but if $R'_j$ is none of the $R_i$, $h_j = O$, then we have

$$A_1, \ldots, A_n[S, k \vdash B \text{ iff } A_1, \ldots, A_n\,[S', k \vdash B.$$

Consider the notion of $(S', k)$-restricted proof so introduced and those $n_i$ corresponding to $R'_i$ not occurring in **L**. By setting these $n_i = 1$, 2, 3, . . ., we get closer and closer approximations to **L′** beginning with **L**.

## 2. Some Examples

Let $R_1$–$R_7$ be respectively the rules UI, EG, TF, Cd, UG, EI, and CQ of Quine's natural deduction system.[2] The notion of a 2-restricted proof for *this* system is captured by the following schemata (where $A'$ is like $A$ except for containing free $x'$ where $A$ contains free $x$):

(i) $(\forall x)A[2 \vdash A'$
(ii) $A'[2 \vdash (\exists x)A$
(iii) If $A \supset B$ is a truth-table tautology, then $A[2 \vdash B$
(iv) $\sim(\forall x)A[2 \vdash (\exists x)\sim A$
(v) $(\exists x)\sim A[2 \vdash \sim(\forall x)A$
(vi) $\sim(\exists x)A[2 \vdash (\forall x)\sim A$
(vii) $(\forall x)\sim A[2 \vdash \sim(\exists x)A$

The definitive character of the 2-restricted system is represented in the result that:

There is a 2-restricted proof of $B$ from the premiss A iff $A[2 \vdash B$ is an instance of one of (i)-(vii).

There are no inference schemata corresponding to UG and EI since no 2-line deduction using either of these is a *finished deduction* (see Quine, *ibid.*, p. 162). (There is no inference schema corresponding to Cd since correct 2-line deductions using Cd have no premisses—a case lying outside the scope of the notions introduced in §1.) It should be noted that it is by no means common that the restricted proofs or consequences of a given system can be so easily schematized.

[2] W. V. Quine, *Methods of Logic*, revised edition (New York, 1959).

So much for an example based on Quine's system; let us consider a different point of departure. Let $R_1$-$R_{13}$ ($m_1$-$m_{13}$) be respectively the rules (integers) reit (2), imp int (1), imp elim (1), conj int (2), conj elim (2), dis int (2), neg elim (1), $neg_2$ int (2), $neg_2$ elim (1), neg conj int (2), neg conj elim (2), neg dis int (2), and neg dis elim (2) of Fitch's system.[3] (So that $S = 2, 1, 1, 2, 2, 2, 1, 2, 1, 2, 2, 2, 2,.$) Rather than attempting to axiomatize some notion of restricted proof or consequence for this system, we simply list some examples of $(S, k)$ and $S$-restricted consequences for various $k$ and then note some general relations among such (for *this* system). Letting $\Gamma$ be an arbitrary string of premisses ($\Gamma$ may be empty in the presence of other premisses), we have:

(1) $\Gamma, A \& B(S, 3 \vdash B \& A$
(2) $\Gamma, A(S, 4 \vdash \sim A \supset B$
(3) $\Gamma, A \vee B(S, 5 \vdash B \vee A$
(4) $\Gamma, A \vee B, \sim A(S, 5 \vdash B.$
(5) If $\Gamma(S, k \vdash A$, then $\Gamma(S \vdash A$
(6) If $\Gamma(S, 1 \vdash \gamma$ and $B, B(S, 1 \vdash \gamma$, then $\Gamma, A \vee B(S, 5 \vdash \gamma$

With respect to (6), it might be noted that the more general

(7) If $\Gamma, A(S, k_1 \vdash \gamma$, and $\Gamma, B(S, k_2 \vdash \gamma$, then
$\Gamma, A \vee B(S, k_1 + k_2 + 3 \vdash \gamma$

does *not* obtain since, although we can find a deduction of $\gamma$ from $\Gamma$, $A \vee B$ given that the antecedent of (7) obtains, we have no guarantee that this (or any) such deduction meets the conditions for $(S, k_1 + k_2 + 3)$ restricted consequence given $S$ as above. For example, both $A(S, 2)(A \vee \gamma) \vee (B \vee \delta)$ and $B(S, 2)(A \vee \gamma) \vee (B \vee \delta)$ obtain, but it is false that $A \vee B(S, 7 \vdash A \vee \gamma) \vee (B \vee \delta)$.

It might be noted that for any system **L** which contains (analogues of) Fitch's rule imp int and imp elim we have the following quasi-deduction-thorem. Let $m_i$ and $m_j$ be the integers in $S$ corresponding to imp int and imp elim respectively, let $S'$ be like $S$ except for having $m'_i = m_i + 1$ where $S$ has $m_i$ and let $S''$ be like $S$ except for having $m'_j = m_j + 1$ where $S$ has $m_j$. Then we have:

(8) If $\Gamma, A[S, k \vdash B$ then $\Gamma[S', k+1 \vdash A \supset B$
(9) If $\Gamma[S, k] A \supset B$ then $\Gamma, A[S'', k+2 \vdash B.$

## 3. Application I: Imperfect Reasoners

Consider a computer which knows several natural deduction rules (perhaps learned at Turing's knee), but which—being "all too human" —can err. Our computer, call it Mycroft, is known to make on the

[3] Frederick B. Fitch, *Symbolic Logic* (New York, 1952).

average one mistaken step in 100 when making deductions, and his errors are independent in the sense that mistakes made before the nth step in a deduction have no bearing on whether the nth step goes wrong. Assume that Mycroft, or Mike for short, knows only *modus ponens* and *modus tollens* and that we want to know the consequences derivable from a set of formulas using these two rules, but are only interested in those formulas that Mike claims as consequences which have at least a .96 chance of being correct. Such a consequence—a "*reliable* consequence"—amounts to a 6-restricted consequence in terms of the machinery introduced in section 1 with **L** = (*modus ponens*, *modus tollens*).[4] (For $[.99]x = .96$, $x \simeq 6$; it would help if Mike were good at logarithms.)

Consider a slightly different situation in which Mike is completely reliable but takes ten times as long to do *modus tollens* as *modus ponens* (say he takes, respectively, one and ten seconds for these operations). Call $B$ a $t$-second consequence of $A_1, \ldots, A_n$ (for Mike) iff Mike can derive $B$ from $A_1, \ldots, A_n$ in $t$ seconds. We can again readily define this in terms of the machinery developed in section 1.

Again, a somewhat similar position would be reached if each time our computer made a certain inferential step it levies some small fee (differing perhaps with the sort of step at issue). Then if we were subject to the constraint of a finite budget this too would circumscribe a limit around the range of the deductive consequences that the computer could make available to us.

## 4. Application II: Epistemic Logic

According to the concept of belief in terms of an implicit commitment to believe, all of the logical consequences of beliefs are themselves believed. This view of belief is sometimes expressed by saying that the believers at issue in a logical theory of the subject are to be logically omniscient and finds its formalization in the meta-principle:

(1) If $B$ is a logical consequence of $A_1 \ldots, A_n$, then $\mathscr{B}xA_1 \,\&\ldots\&\, \mathscr{B}xA_n \supset \mathscr{B}xB$ is a theorem.

The logic of this conception of belief has been amply investigated by Hintikka.[5] According to a second and more restricted construction, one believes all and only those things to which one would be prepared to give explicit assent if questioned. The logic of this concept has not

[4] It is to be assumed for present purposes that we can ask Mike to carry through any particular deduction but once. If re-checking were possible the situation would be altered.

[5] See Jaakko Hintikka, *Knowledge and Belief* (Ithaca, 1962).

been much studied by philosophical logicians and, indeed, it seems doubtful that this concept can have an "interesting" logic. (It is readily conceivable, for example, that a person can believe $A \& B$ in this sense, and yet fail to believe $A \vee C$.) With this *overt* sense of belief, we must be prepared to encounter believers who are "logically blind."

It seems desirable to explore the possibility of a middle course between these extremes, to investigate a belief construction such that, to put it metaphorically, believers are neither logically omniscient nor logically blind but rather have a logical vision of limited range. Thus in an earlier publication[6] I had proposed a system of epistemic logic for this kind of construction in which (1) is replaced by the weaker:

(2) If $B$ is an *obvious* consequence of $A_1, \ldots, A_n$, then $\mathscr{B}xA_1 \& \ldots \& \mathscr{B}xA_n \supset \mathscr{B}xB$ is a theorem,

where the following rough characterization of "obvious consequence" is given: $B$ is said to be an obvious consequence of $A_1, \ldots, A_n$ if $B$ is deducible from $A_1, \ldots, A_n$ in some small number of inferential steps. It is further specified that (2) is not to be used more than once in any proof or deduction. It was observed that this notion of obvious consequence has some similarity to Hintikka's notion of a "surface tautology."[7]

It is to be presumed that a rigorous characterization of the notion of an "obvious consequence" can be given in terms of the machinery developed in Section 1 above. Given such an explication, the interpretation of (2) becomes well-specified and definite. On the other hand, we are now in a position to make finer distinctions than this construction allows.

Although we are not prepared to give a complete characterization of the notion of obvious consequence, it may be helpful to make some brief remarks on the subject. We assume that the obvious consequence relation is *roughly* a variety of restricted consequence or proof, but one might well want to make additional requirements on the *kind* of deductions that can result in an *obvious* consequence. For example, if the underlying system is one which allows subordinate proofs, one presumably would also want to require that these themselves be (in some yet more stringent sense) "obvious." This might be thought to be especially appropriate where the subordinate proof involves a *reductio* argument. Thus one might be led to modify the notion of

[6] Nicholas Rescher, "The Logic of Belief Statements," *Topics in Philosophical Logic* (Dordrecht, 1968), pp. 40–53.

[7] Cf. Jaakko Hintikka, " 'Knowing Oneself' and Other Problems in Epistemic Logic," *Theoria*, vol. 32 (1966), pp. 1–13.

restricted consequence as follows (for simplicity we consider only the case where all subordinate proofs are immediately subordinate to the main proof):

> Let $S$, the $R_i$, and $k$ be as before. Let $S' = <0_1, \ldots, 0_n>$ where $0_1 \leqslant m_1, \ldots, 0_n \leqslant m_n$, and let $k' < k$. Then an $(S, S', k, k')$-*restricted proof* of $B$ from $A_1, \ldots, A_n$ is one that satisfies (i) and (ii) as before, and (iii) is such that in any subordinate proof no rule $R_i$ is applied more than $0_i$ times, and (iv) no subordinate proof is longer than $k'$ lines in length.

For example, take the system of Fitch considered in Section 2. Letting $0_i = m_i - 1$ and $k' = k - 1$, we have as analogues of (1), (3), and (5) of that section:

> (1′) $\Gamma, A \& B\ (S, S', 3, 2 \vdash B \& A$
> (3′) $\Gamma, A \vee B\ (S, S', 5, 4 \vdash B \vee A$
> (5′) If $\Gamma\ (S, S', k, k' \vdash A$, then $(S, k \vdash A$.

Given a notion of obvious consequence, one could then define the following concepts: $A$ is *patently valid* iff $A$ is an obvious consequence of $B \vee \sim B$ for some formula $B$; $A$ is *patently inconsistent* iff for some formula $B$, $B \& \sim B$ is an obvious consequence of $A$.

Assume that certain beliefs are in some appropriate sense epistemically *basic*. In this case we write: $\mathscr{B}_0 xA$. We define $\mathscr{B}_{i+1} xB$ so as to obtain iff $B$ is an obvious consequence of $A_1, \ldots, A_k$ where $\mathscr{B}_i xA_1, \ldots, \mathscr{B}_i xA_k$. Then, instead of (2), we can adopt the meta-principle:

> (3) If $B$ is an *obvious consequence* of $A_1, \ldots, A_n$, then $\mathscr{B}_{m_1} xA_1 \& \ldots \& \mathscr{B}_{m_n} xA_n \supset \mathscr{B}_{\max(m_i)+1} xB$ is a theorem.

This embraces the special case:

> (4) If $B$ is an obvious consequence of $A$, then $\mathscr{B}_n xA \supset \mathscr{B}_{n+1} xB$.

If the relation of obvious consequence allows repetition, i.e., if for all $i$ such that $1 \leqslant i \leqslant n$, $A_i$ is an obvious consequence of $A_1, \ldots, A_n$, then (3) follows from the definition of $B_n xA$.

Such a conception of obvious consequencehood clearly has a worthwhile role to play in epistemic logic. For if $\mathscr{E}$ is an epistemic operator representing a cognitive relationship such as knowledge ($\mathscr{K}$) or belief ($\mathscr{B}$), then it would not be unproblematically plausible to base the logic of this relation on the inferential rule:

$$A_1, A_2, \ldots, A_n \vdash B$$

$$\frac{\mathscr{E}xA_1, \mathscr{E}xA_2, \ldots, \mathscr{E}xA_n}{\therefore\ \mathscr{E}xB}$$

This rule in effect postulates logical omniscience: cognizers are supposed to cognize all logical consequences of their cognitions; and so, for example, in cognizing the axioms they immediately cognize all the theorems. This is all very well as long as we take the stance that our cognizers are logical systems or idealized "fully rational intellects," but is patently inappropriate when imperfect humans are at issue. Here it seems altogether plausible to assume a limited range of logical vision, a sort of epistemic near-sightedness. Of course, any reasonably rational flesh-and-blood person can be assumed to make certain obvious inferences—e.g., if he believes (or knows) the conjunction $A$ & $B$ he will, of course, also believe (or know) conjuncts $A$ and $B$. The "obvious" consequences of what he cognizes may, of course, also be supposed to be cognized by him. Yet his inferential capacities are not unlimited, but finite, and we must recognize that he can carry out inferences only "up to a point." He is affected by a sort of inferential myopia, and this is something that *we* must recognize before proceeding in our epistemic logic system to stipulate that the inferential consequences of what is cognized will also be cognized themselves.

## 5. Conclusion

The idea of a restricted inference seems to offer a rigorous way of introducing into logic a conception that seemingly has no place here: the economists' concept of limited resources, of a finiteness of means, in short, of *scarcity*. In ways we have attempted to define and illustrate, the mechanisms of restricted inference provide a basis for injecting into considerations of logical deduction the operation of the familiar restrictive limitations of human finitude: limitations forced upon us in situations of limited time, accuracy, or logical acumen. The implications of such limitations may well repay further study—they are unquestionably of *practical* importance and may be presumed to have substantial *theoretical* interest as well.[8]

[8] This essay is an expanded version of a paper written in collaboration with Zane Parks and first published in *Logique et Analyse*, vol. 55 (1971), pp. 675–683.

IX.

# Modal Elaborations of Propositional Logics and Their Epistemic Aspect

## 1. Modally Augmented Systems

ONE interesting perspective upon modal logic is obtained by beginning with a nonmodal system, and then developing a modal system "around" it, so to speak, by construing necessity in the "surrounding" modal system as provability within the initial system.[1] Modality, so conceived, is obtained in the broader system by bridging rules linking the necessity operator in this system to thesishood at the nonmodal strating point. The aim of the present paper is to trace out one line of thought along which this idea can be implemented.

Let **L** be an arbitrary system of (nonmodal) propositional logic based upon negation ($\neg$), conjunction (&), and implication ($\rightarrow$) as propositional operators. The theses of **L** are to be derived from certain (at this point unspecified) axioms by the rules of substitution and *modus ponens*. (We shall write $\vdash_{X} A$ to indicate that $A$ is a thesis of the system X.)

To obtain the modal system **ML**, the modal augmentation of the initial system **L**, we introduce the modal operator of necessity ($\Box$) subject to the rules and axioms of the following sort:

I. *Modal Axioms Internal to* **ML**

(A1) $\vdash_{ML} \Box p \rightarrow p$

(A2) $\vdash_{ML} \Box (p \rightarrow q) \rightarrow (\Box p \rightarrow \Box q)$

II. **ML**-*Internal Rules*

(R1) *Substitution*

(R2) *Modus Ponens*

(R3) *Qualified Necessitation*: *If* $\vdash_{ML} A$, *then* $\vdash_{ML} \Box A$, provided *A is not modal-free*

[1] The origins of this line of thought may be sought in Kurt Gödel, "Eine Interpretation des intuitionstischen Aussagenkalküls," *Ergebnisse eines mathematischen Kolloquiums*, vol. 4 (for 1931–1932, published 1933), pp. 39–40. Here necessity is identified with provability in a certain system. See also Rudolf Carnap, *The Logical Syntax of Language* (London, 1967), pp. 233–260; and W. V. Quine, "Three Grades of Modal Involvement," *Proceedings of the XIth International Congress of Philosophy* (Brussels, 1953), vol. 14, pp. 65–81.

III. **L/ML** *Bridging Rule*

(B) *If* $\vdash_{L} A$, then $\vdash_{ML} \Box A$

IV. *Metarule of Closure*

(C) $\vdash_{ML} A$ *only if A's being a thesis of* **ML** *follows from the preceding rules and axioms.*

It is readily shown that:

*If A is a thesis of* **L**, *then A is a thesis of* **ML**.

For if $A$ is a thesis of **L**, then by (B), $\vdash \Box A$ in **ML**, which by (A1) and (R2) yields $A$. Moreover:

*If A is a thesis of* **ML** *and A is modal-free, then A is a thesis of* **L**.

For suppose $A$ is a thesis of **ML** and $A$ is modal-free. Then by (C) it follows, in view of the nature of the rules and axioms, that $A$ could be the result only of (A1) and (R2). Hence we must have $\Box A$ as a thesis of **ML**. But this in turn could only be the result of rule (B). Hence $A$ is a thesis of **L**. It follows from these findings that **ML** must be a conservative extension of **L** (whenever **ML** is consistent[2]),[3]

One obvious consequence of this approach is represented by the theorem:

*If* **L** = PC, *the classical propositional calculus, then* **ML** = **T**, *the well-known system of Feys- von Wright.*

In one well-known axiomatization, **T** is based on $\Box$ as primitive, subject to the following rules and axioms:

*Rules*

(TI) *If* $\vdash_{PC} A$, *then* $\vdash_{T} A$

(TII) *If* $\vdash_{T} A$, *then* $\vdash_{T} \Box A$

*Axioms*:

(Ti) $\vdash_{T} \Box p \supset p$

(Tii) $\vdash_{T} \Box (p \supset q) \supset (\Box p \supset \Box q)$

Throughout, when **L** = PC, we shall write $\supset$ for $\rightarrow$. Now given (R2), (TI) follows from (A1) and (B). Moreover: (Ti) = (A1), and (Tii) = (A2). Thus **ML** is at least as strong as T when **L** = PC. It can also be shown (though we shall not do so here), that it is no stronger, so that

[2] The proof that **ML** is a conservative extension of **L** depends critically on the use of the rule of closure (C), and such a use is possible only on the assumption that **ML** is consistent.

[3] The system **L′**, is an extension of **L** if the vocabulary of **L′** contains that of **L** as a subset and every **L**-thesis is an **L′**-thesis. **L′** is a conservative extension of **L** if it is an extension such that all **L′** theses formulated in the **L** vocabulary will also be theses of **L′**.

**ML** = **T** in this case. Finally, (TII) is proved as follows: If $A$ is not modal-free and a thesis of **ML** then $\Box A$ follows by (R3). If $A$ is modal-free and a thesis of **ML** then, using (C), we note that $A$ could come only from (A1) and (R2), and so again we must have $\Box A$ as a thesis of **ML**.

Correspondingly, it is also readily seen that if we make a sufficient addition to Category I of **ML**-internal rules (by adding $\vdash_{\mathbf{ML}}\Box A_1$, $\vdash_{\mathbf{ML}}\Box A_2, \ldots, \vdash_{\mathbf{ML}}\Box A_n$, where $A_1, A_2, \ldots, A_n$ are any syntactically suitable set of PC-axioms) then the modally augmented system **ML** will have to contain **T** (no assumptions whatsoever being made about the initial system **L**).

Suppose now that (R3) were dropped from the construction procedure for **ML** and that (A2) were strengthened to:

(A3) $\vdash_{\mathbf{ML}}\Box(p \supset q) \supset \Box(\Box p \supset \Box q)$.

We now have the result:[4]

*If* **L** = PC, *then* **ML** = S3.

In the face of this finding, it is quite readily established that **S4** can also be developed as a modally augmented system by the reinstatement of (R3), and that **S5** will then be obtained by further addition of the axiom:

(A4) $\vdash_{\mathbf{ML}}\Diamond p \supset \Box\Diamond p$ *where* $\Diamond q = \neg\Box\neg q$

These observations indicate how systems of modal logic can be developed, via bridging rules from arbitrary systems of nonmodal propositional logic, in such a way that in the case of classical propositional calculus (PC) as the starting-system we obtain the spectrum of the most familiar modal systems. This suggests the potentially interesting question, or question family, of the modally augmented systems resulting from initial propositional logics *weaker* than PC, such as intuitionistic propositional logic. On the other hand, if the initial system is very strong—and specifically is a system of *arithmetic* rather than propositional logic—then a consistent modal augmentation becomes impossible when the modal system is strong enough to contain the thesis: $\Box(\Box p \rightarrow p)$.[5] Since **ML** has (A1) as one of its axioms, and also the necessitation rule, clearly the system will have $\Box(\Box p \rightarrow p)$; hence modal augmentation along the lines indicated above is impossible when **L** is so strong a system. But note that the proof that a general necessitation rule is derivable in **ML** depends

[4] This is obvious by inspection of the development of **S3** given in G. E. Hughes and M. J. Cresswell, *An Introduction to Modal Logic* (London, 1968), p. 344.

[5] See footnote 6 below, as well as its context in the text.

on (R3). In the following we shall construct a sequence of modally augmented systems which lack (R3), and do not have the necessitation rule, but only some highly restricted form thereof.

## 2. Minimality

In considering modal systems as modal augmentations of non-modal systems, we can thus regard the derived modal systems as developed (by a suitable procedure) from the underlying logical system **L**. This leads to the problem of what modal systems are associated with a logical system, and to the question of when a modal system of the type of **ML**, satisfying the bridging-rule (B), could be properly called (in any sense) a modal system *for* **L**. The following two conditions seem minimally necessary:

(i) **ML** is consistent if **L** is so
(ii) **ML** is a conservative extension of **L**, so that if $A$ is a thesis of **ML** and $A$ is modal free (m.f.) then $A$ is a thesis of **L**.

This is a very weak sort of minimality; thus when **L** = PC, then all of the Lewis systems will satisfy these conditions. However, for the present, we shall characterize any system satisfying these conditions as a *minimally adequate modal augmentation of* **L**.

If the initial system **L** is complete (in the strong sense that for every $A$ in its vocabulary either $A$ or $\neg A$ is a thesis of **L**), and **ML** (constructed by the use of the bridging rule (B)) is consistent, then **ML** must satisfy the remaining minimality condition if it has (A1) and (R2). For let $A$ be (m.f.) and suppose $A$ is a thesis of **ML** but not of **L**. Then, since **L** is complete, $\neg A$ in **L** and by (B) $\Box\neg A$ in **ML**. This by (A1) and (R2) yields $\neg A$ in **ML**, contradicting the consistency of **ML**.

However, (A1) and (R2) are not sufficient to assure that **ML** satisfies the minimality property in the more general case when **L** is incomplete. But in this general case we have the theorem:

*The modally augmented system* **ML**, *satisfying* (B), (*A*1), *and* (R2), *is a conservative extension of* **L** *if and only if it also satisfies the two following rules.*

(R3′) *If* $\vdash_{\mathbf{ML}} A$ *then* $\vdash_{\mathbf{ML}} \Box A$, *where A is m.f.*
(B′) *If* $\vdash_{\mathbf{ML}} \Box A$ *then* $\vdash_{\mathbf{L}} A$, *where A is m.f.*

*Proof*: If **ML** is a conservative extension of **L**, suppose $A$ is m.f. and a thesis of **ML**. Then $A$ in **L**, and hence by (B) $\Box A$ in **ML**, so that (R3′) obtains. And if $\Box A$ in **ML** where $A$ is m.f., then by (A1) and (R2) $A$ in **ML**, which by assumption yields $A$ in **L**. Hence (B′) also holds. To prove the converse, suppose that **ML** satisfies both (R3′)

and (B′). If $A$ is m.f. and a thesis of **ML** then by (R3′) $\Box A$ in **ML**, which by (B′) yields $A$ in **L**.

Note that in the previous section we took **ML** as satisfying (A1), (A2), (R1), (R2), (B) and the metarule (C). We now realize that all that was needed in order to show that **ML** is a conservative extension of **L** apart from (B), (A1), and (R2) was the metarule (C). Hence (C) entails (R3′) and (B′) under these conditions.

## 3. A Modal Hierarchy

As a result of the foregoing, in order to satisfy the minimality requirement, a system **ML** has to have a restricted form of the necessitation rule (R3′), but need not have the general unrestricted rule, (R3). For if **ML** is minimal, and has $\Box A$ as a thesis then it does *not* follow that $\Box\,\Box A$ is a thesis in **ML**. This suggests that we can construct yet another modal system "around" the minimal modal system **ML**, by construing necessity in it as provability in **ML**. As we shall see, this yields a hierarchy of modal systems, each a conservative extension of the previous one. The system which is the "least upper bound" of the hierarchy has the rule of necessitation in it, and hence if we add to it (A2) and (R1), it becomes **T**-like in the sense that when **L** = PC the system is **T**.

In order to show all this, we shall prove it for systems assumed to have (A2) and (R2), but since the proof does not make use of these, the corresponding general assertion about minimal systems follows. Moreover, the consistency (supposing that **L** is a consistent system of propositional logic) of each of the modal systems of the hierarchy follows immediately from the consistency of **T**, since they must all be fragments of **T**.

To facilitate the discussion, we shall introduce the notion of the *degree of a formula*.

1. A formula $A$ (of **ML**) is of degree 0, if it is modal free. We shall write this as $\deg(A) = 0$.
2. If $A$ has the form $\Box B$, and $\deg(B) = n$ then $\deg(A) = n + 1$.
3. $\deg(\neg A) = \deg(A)$.
4. $\deg(A \,\&\, B) = \max(\deg(A), \deg(B))$.
5. $\deg(A \rightarrow B) = \max(\deg(A), \deg(B))$.

The modal system **ML** is said to be of degree $n$, (in this case we shall say **ML** = $\mathbf{M}_n\mathbf{L}$) if it has theses of degree $n$ but none of greater degree, so that $\vdash_{\mathbf{M}_n\mathbf{L}} A$ implies that $\deg(A) \leq n$. We allow $n$ to be any non-negative integer, understanding $\mathbf{M}_0\mathbf{L}$ to be the nonmodal system **L**.

Given the nonmodal system $\mathbf{L} = \mathbf{M}_0\mathbf{L}$, we can construct simultaneously by induction on $n$ ($n = 1, 2, \ldots$) a hierarchy of systems where each is the modal augmentation of the preceding one; thus:

I. *Modal Axioms Internal to* $\mathbf{M}_n\mathbf{L}$

(A$_n$1) $\vdash_{\mathbf{M}_n\mathbf{L}} \Box p \to p$

(A$_n$2) $\vdash_{\mathbf{M}_n\mathbf{L}} \Box (p \to q) \to (\Box p \to \Box q)$

II. *$M_nL$ Internal Rules*

(R$_n$1) *Substitution*

(R$_n$2) *Modus Ponens*

III. *Bridging Rules*

(B$_n$) *If* $\vdash_{\mathbf{M}_{n-1}\mathbf{L}} A$ *then* $\vdash_{\mathbf{M}_n\mathbf{L}} \Box A$.

A system $\mathbf{M}_n\mathbf{L}$ satisfying these rules and axioms is not in general a conservative extension of $\mathbf{M}_{n-1}\mathbf{L}$. However, by a similar proof to that given for the basic case, one can prove (by induction on $n$) that $\mathbf{M}_n\mathbf{L}$ is a conservative extension of $\mathbf{M}_{n-1}\mathbf{L}$ if and only if it satisfies the following:

(R$_n$3) *If* $\vdash_{\mathbf{M}_n\mathbf{L}} A$ *then* $\vdash_{\mathbf{M}_n\mathbf{L}} \Box A$, *where* $\deg(A) \leq n - 1$

(B$'_n$) *If* $\vdash_{\mathbf{M}_n\mathbf{L}} \Box A$ *then* $\vdash_{\mathbf{M}_{n-1}\mathbf{L}} A$, *where* $\deg(A) \leq n - 1$

Still, there is nothing to assure us that $\mathbf{M}_n\mathbf{L}$ is a system of degree $n$, i.e., all its theses have degree $\leq n$. This effect may be introduced by a restriction on the language of $\mathbf{M}_n\mathbf{L}$, allowing it to have as well-formed formulas (wffs) only those whose degree is not greater than $n$.

A possible motivation for such an approach is the following. Just as in the basic case, we consider a formula with modalities as foreign to the non-modal system **L**, because the language of **L** does not have the degree one necessity operator, so in the general case: if $A$ is a formula itself of degree $n$, then $\Box A$ is not a meaningful formula of $\mathbf{M}_n\mathbf{L}$ because (although we here use the same necessity symbol) this additional modality represents a *new*, deeper modal operator foreign to the language of $\mathbf{M}_n\mathbf{L}$. For example, one might want to claim that if $A$ is a thesis of a logical system **L** then it is logically necessary, so that $\Box A$ is a thesis of a modal system, but that $\Box\Box A$ is meaningless (in the sense of undefined).[6] Alternatively, one might consider this iteratedly modal formula as meaningful, but hold that the first occurrence of the necessity operator has a meaning different from

[6] On this approach, modalities are to be viewed as classifiers of sentences (or as predicates thereof) rather than as sentential operators like negation. Correspondingly, modal iteration makes no sense. Thus if necessity is construed as provability "is provably provable" could—from a suitable perspective—be viewed as a nonsense.

that of the second one, so that one should properly write it as $\Box_2\Box_1 A$ rather than simply $\Box\Box A$.

Thus, if the initial system **L** were of a very strong sort (specifically, a system of arithmetic), then the augmented modal systems (if they are to be plausible as modal systems) should be cut off at some finite level.[7] Alternatively, this result could be evaded by construing iterated modalities as equivocal, along the lines of the present proposal. This suggests also that we can consider modal augmentation of modal systems, i.e., where **L** is itself a modal system. Consider for example the case where **L** = PC. From it we construct $\mathbf{M_1L}$ as before, only instead of using the symbol '$\Box$' we write '$\Box_1$'. Thus $\Box_1 A$ in $\mathbf{M_1L}$ if and only if $A$ in PC. Now we add to $\mathbf{M_1L}$ the symbols and axioms of arithmetic and considering this system as our base we construct its modal augmentation, using this time the symbol '$\Box_2$'. Now in this system, call it $\mathbf{M_2L}$, $\Box_1 A$ means $A$ is a PC-thesis, $\Box_2 A$ means $A$ is either a thesis of PC or a thesis of $\mathbf{M_1L}$ or a thesis of $\mathbf{M_1L}$ + arithmetic. In short, since the latter is an extension of the first two systems, it is a thesis of $\mathbf{M_1L}$ + arithmetic. Clearly in $\mathbf{M_2L}$ no formula of the form $\Box_1\Box_2 A$ is a thesis, since no rule allows construction of it. Moreover, for the same reason, $\mathbf{M_2L}$ does not have $\Box_2\Box_2 A$, which yields the inconsistency indicated by Montague.

Yet another way of assuring ourselves that the system $\mathbf{M_nL}$ will be of degree $n$ is by introducing the following:

*Metarule of Closure*

($C_n$) $\vdash_{\mathbf{M_nL}} A$ *only if A's being a thesis of* $\mathbf{M_nL}$ *follows from the preceding rules and axioms* (*of groups* (I), (II), *and* (III))

It can be easily shown that by adding this rule to the system, $\mathbf{M_nL}$ is a conservative extension of $\mathbf{M_{n-1}L}$. (The proof is similar to the one given in the first section.) Thus in order that our system will satisfy the minimality condition it suffices that we introduce ($C_n$), and then ($R_n$3) and ($B'_n$) will be derived rules of the system.

By the construction of the hierarchy, each system is an extension of the preceding one. Since each system allows as theses formulas of degree $n$ or less, an "upper bound" of this set will have the same rules and axioms as $\mathbf{M_nL}$, except for the bridging-rule and the metarule of closure. If we call this system $\mathbf{M^*L}$, it transpires that this system may be defined by the stipulation that

$\vdash_{\mathbf{M^*L}} A$ *is to obtain just in case there exists an n such that* $\vdash_{\mathbf{M_nL}} A$.

[7] See Richard Montague, "Syntactical Treatments of Modality," *Proceedings of a Colloquium on Modal and Many-valued Logics, Acta Philosophica Fennica*, fasc. 16 (1963), pp. 153–167.

Now clearly, **M*L** will have the rule of necessitation. For if $A$ is a thesis of **M*L**, then there will be an $n$ such that $A$ is a thesis in $\mathbf{M}_n\mathbf{L}$, and hence $\Box A$ in $\mathbf{M}_{n+1}\mathbf{L}$, so that $\Box A$ in **M*L**. Moreover, **M*L** will clearly have (A1) (A2), (R1), (R2), and hence **M*L**, thus defined, will be an extension of **ML**, the modal augmentation of **L** introduced in the previous section. But **ML** is an extension of each of the $\mathbf{M}_n\mathbf{L}$, and therefore **M*L** = **ML**. From this it follows immediately that when **L** = PC, then **M*L** = **T**.[8]

## 4. Application to Epistemic Logic

One particularly interesting application of this machinery arises when the modal operator at issue in the modal system that "surrounds" the basic system in view represents an *epistemic* operator $K$ (for "it is known that"). Here one takes the view that the "knower" is, in effect, itself a logical system whose cognitive target or object is a certain logical system (or a person who is duly omniscient with respect to that system).

This point of view brings the considerations of the present chapter into alignment with those considered in our earlier discussion of epistemic logic (in Chapter VI above). This parallelism is indicated in the following tabulation:

| | *Epistemic Principle* (cf. pp. 100–104 above) | | *Analogies in the Present Discussion* |
|---|---|---|---|
| (K1) | $\vdash Kxp \supset p$ | (A1) | $\vdash_{\mathbf{ML}} \Box p \supset p$ |
| (K2) | $\vdash Kx(p \supset q) \supset (Kxp \supset Kxq)$ | (A2) | $\vdash_{\mathbf{ML}} \Box(p \supset q) \supset (\Box p \supset \Box q)$ |
| (K3) | $\vdash_{\mathbf{L}} A \Rightarrow \vdash KxA$ | (B) | $\vdash_{\mathbf{L}} A \Rightarrow \vdash_{\mathbf{ML}} \Box A$ |
| (K′) | $\vdash A \supset B \Rightarrow \vdash KxA \supset KxB$ | (T) | $\vdash_{\mathbf{ML}} A \supset B \Rightarrow \vdash_{\mathbf{ML}} \Box A \supset \Box B$ |
| (K) | $\vdash A \Rightarrow \vdash KxA$ | (R3*) | $\vdash_{\mathbf{ML}} A \Rightarrow \vdash_{\mathbf{ML}} \Box A$ |
| (K4) | $\vdash Kxp \supset Kx(Kxp)$ | | $\vdash_{\mathbf{ML}} \Box A \supset \Box\Box A$ |
| (K5) | $\vdash \sim Kxp \supset Kx(\sim Kxp)$ | | $\vdash_{\mathbf{ML}} \sim\Box A \supset \Box\sim\Box A$ |

In these circumstances, the validation of the basic principles is relatively straightforward. Let us test a few cases. The thesis

[8] This incidentally exhibits (and the fact may be obvious enough from other points of view) how there is an infinite sequence of distinct modal systems intermediate between the straightforward modalization of PC (viz., the system $\mathbf{M}_1\mathbf{L}$ with **L** = PC), and the system **T**. This suggests a further question: It has been shown that if we successively add to the system **T** axioms of the form $\Box^n p \supset \Box^{n+1} p$ $(n = 1, 2, \ldots)$ then the result is a descending sequence of systems $\mathbf{T}_1, \ldots, \mathbf{T}_n, \ldots$ between S4 and **T** with $\mathbf{T}_1 = \mathbf{T} + (\Box p \supset \Box\Box p) = $ S4, and each $\mathbf{T}_{n+1}$ a proper subsystem of $\mathbf{T}_n$. (This is an unpublished result of B. Sobocinski: see Ivo Thomas, "Modal Systems in the Neighborhood of T," *Notre Dame Journal of Formal Logic*, vol. 5 (1964), p. 59.

(A1) $\vdash_{\mathbf{K}} Kp \to p$

is an obvious principle of epistemic logic. And the same holds for

(A2) $\vdash_{\mathbf{K}} K(p \to q) \to (Kp \to Kq)$

The bridging rule

(B) If $\vdash_{\mathbf{L}} A$, then $\vdash_{\mathbf{K}} KA$

implements the idea that the **K**-system is "logically omniscient" with respect to the demonstrable theses of the basic target-system **L**. Rules (R1)–(R2) warrant no special notice. However, the rule of qualified necessitation now acquires a particular interest. For this rule says:

If $\vdash_{\mathbf{K}} A$, then $\vdash_{\mathbf{K}} \Box A$, provided $A$ is not modal free

This leads to such consequences as

If $\vdash_{\mathbf{K}} KA$, then $\vdash_{\mathbf{K}} KKA$

If $\vdash_{\mathbf{K}} {\sim} KA$, then $\vdash_{\mathbf{K}} K{\sim} KA$

The rationale of principles of this sort was considered in Chapter VI above. Their inclusion in the system in view indicates that the "knowledge" now at issue calls for "logical omniscience" of a very strong sort indeed. Of course, this circumstance is not as far-fetched as it might otherwise seem to be, since the "knower" with whom we have to deal is, in effect, a logical system, so that "knowledge" now represents not what some finite intelligence in fact knows, but rather what is in principle *knowable* on the basis of certain fundamental assumptions.[9]

[9] This essay is an expanded version of a paper entitled "Modal Elaborations of Propositional Logic" first published in the *Notre Dame Journal of Formal Logic*, vol. 13 (1972), pp. 323–330, and written in collaboration with Ruth Manor.

# SUBJECT INDEX

# INDEX OF NAMES